Wastewater Microbiology:
A Handbook for Operators

Wastewater Microbiology: A Handbook for Operators

Second Edition

Toni Glymph-Martin

American Water Works Association

Disclaimer
The authors, contributors, and publisher do not assume responsibility for the validity of the content or any consequences of their use. In no event will AWWA be liable for direct, indirect, special, incidental, or consequential damages arising out of the use of information presented in this book. In particular, AWWA will not be responsible for any costs, including, but not limited to, those incurred as a result of lost revenue. In no event shall AWWA's liability exceed the amount paid for the purchase of this book.

ISBN: 978-1-64717-246-6   ISBN, electronic: 978-1-61300-769-3
https://doi.org/10.12999/AWWA.20563ed2

Sr. Manager — Product Acquisition & Development: Geoffrey S. Shideler
Manager — Publishing Operations: Gillian Wink
Specialist — Copyright and Permissions: Peggy Tyler
Director of Publishing: John Fedor
Technical Editor: Wiley
Technical Editing Review: Suzanne Snyder
Cover Design and Technical Illustrations: Michael Labruyere
Production: Innodata
Cover Images: zffoto/Shutterstock.com, Toni Glymph-Martin

Library of Congress Cataloging-in-Publication Data
CIP data pending from the Library of Congress

American
Water Works
Association

Contents

To access the online supplementary material,
please visit awwa.org/ww-micro-online.

LIST OF FIGURES

LIST OF TABLES

Preface

Nations are facing many issues, including changing climates, increased demand for our limited natural resources, the dominance of invasive species, and the extinction of natural flora and fauna. The management of water and wastewater is no exception. The water and wastewater treatment industry is changing and ever evolving. Over the years, different methods have been used to treat wastewater. Ancient methods of waste disposal included simply burying waste in holes in the ground. Although we have come a long way since then, it is hard to imagine that there are places where this is still the practice. As recent as 2012, I visited a small community located about 100 miles outside of Haiti in the Dominican Republic. There were no toilets, no running water, and no power. The villagers were still burying waste in holes in the ground. Water from a small creek was used for drinking, bathing, and washing clothes. Many of the children were suffering from parasite infections from drinking untreated water.

Untreated wastewater, composed primarily of biodegradable organic compounds, is easily treated using biological methods. However, the composition of wastewater has also evolved. Once simple, domestic wastewater now contains a variety of man-made chemicals, pharmaceuticals, and complex organic and inorganic compounds. Treatment for these will require more advanced physical and chemical methods. New pollution problems have placed additional burdens on wastewater treatment systems. Today's pollutants, such as heavy metals, chemical compounds, and toxic substances, are more difficult to remove from water. Rising demands on the water supply only aggravate the problem. The increasing need to reuse water calls for better wastewater treatment. These challenges are being met through better methods of removing pollutants at treatment plants or through prevention of pollution at the source. Pretreatment of industrial waste, for example, removes many troublesome pollutants at the beginning, not the end, of the pipeline. To return more usable water to receiving lakes and streams, new methods for removing pollutants are being developed. Advanced waste treatment techniques in use or under development range from biological treatment capable of removing nitrogen and phosphorus to physical–chemical separation techniques such filtration, carbon adsorption, distillation, and reverse osmosis. These wastewater treatment processes, alone or in combination, can achieve almost any degree of pollution control desired. Waste effluents purified by such treatment can be used for industrial, agricultural, or recreational purposes, or even drinking water supplies.

Technologies for treating wastewater have evolved. However, whether methods are simple or complex, microorganisms play a dominant role in removing organic pollutants from wastewater.

REGULATIONS

Wastewater treatment regulations are established to protect public health and the environment by ensuring that wastewater is treated to remove harmful contaminants before it is discharged into water bodies or reused. In the United States, the Clean Water Act (CWA) is the primary federal law governing water pollution. It aims to restore and maintain the chemical, physical, and biological integrity of the nation's waters. Under the CWA, the National Pollutant Discharge Elimination System (NPDES) program requires wastewater treatment facilities to obtain permits for discharging treated effluent into surface waters (USEPA 2024). These permits specify the types and amounts of pollutants that can be discharged.

The US Environmental Protection Agency (USEPA) sets national effluent guidelines for municipalities and industries that discharge wastewater. These guidelines establish limits on the amounts of specific pollutants that can be released. Wastewater treatment facilities must obtain permits that regulate the discharge of treated effluent. These permits often include specific pollutant limits, monitoring requirements, and reporting obligations. Facilities are also required to monitor their discharges regularly and report the results to regulatory authorities. This ensures compliance with permit conditions and helps with tracking environmental impacts. Regulatory agencies have the authority to enforce compliance through inspections, fines, and other penalties for violations. This ensures that facilities adhere to the legal requirements. Many regulations include provision for public involvement, such as public notice and comment periods for permit applications, to ensure transparency and community involvement in decision-making processes.

Wastewater treatment regulations are critical for protecting water resources and public health. They involve a combination of permitting, setting standards, monitoring, enforcement, and public participation. While specific regulations vary by region, the overarching goals are similar: to minimize pollution and promote sustainable water management.

REFERENCE

USEPA (US Environmental Protection Agency). 2024. "Summary of the Clean Water Act. 33 U.S.C. §1251 et seq. (1972)." www.epa.gov/laws-regulations/summary-clean-water-act (accessed November 15, 2024).

Acknowledgments

First and foremost, to God be the Glory! To my husband Clarence, who keeps me laughing, thank you for pursuing me and never giving up. Laughter is like medicine. To my sons, Lawrence and Daniel, and their beautiful wives, Kayla and Noemi, you have given me purpose and joy. To my sister and colleague, Dr. Lynetta Davis, I know you don't think you should be acknowledged, but you have no idea how much your friendship has helped to keep me sane. Most of all, to all of the wastewater treatment operators I have encountered in the 45 years I have worked in this field, I am proud of what we do. Rarely do you get the credit you deserve for the vital work that you do. Although you are often considered a pollution source for discharging treated effluent into the waterways, you are not the villains. You are the heroes. You take 100% of the waste from homes, industry, and stormwater and reduce it by 99%. I would hate to imagine what the world would be like without you! Keep up the good work. I am honored to work with you.

Chapter 1

Wastewater Treatment Overview

One of the most common forms of pollution control in the United States is the treatment of wastewater. Wastewater is treated through a vast system of collection sewers, pumping stations, and water treatment facilities. Sewers collect wastewater from homes, businesses, and many industries and deliver it to wastewater facilities for treatment. Most treatment facilities are designed to clean wastewater for discharge into streams or other receiving waters, or for reuse. Years ago, when sewage was dumped into waterways, a natural process of purification began. First, the sheer volume of clean water in the stream diluted most of the waste. Bacteria and other small organisms in the water consumed organic matter in wastewater. Today's higher populations and greater volume of domestic and industrial wastewater require that communities give nature a helping hand. The basic function of wastewater treatment is to speed up the natural processes by which water is treated.

Raw wastewater, or "sewage" collected from a community, flows through a collection system network into the wastewater treatment facility. Wastewater can come from several sources:

- Domestic wastewater from homes contains waste streams from toilets, wash water, cooking, and bathing.
- Commercial wastewater comes from small businesses such as restaurants, schools, hospitals, and laundromats.
- Industrial wastewater is generated mostly from manufacturing facilities.
- Stormwater is collected during rain events.
- Groundwater enters the collection system through faulty sewer joints or cracked pipes, generally termed "infiltration and inflow."

There are also different types of sewer systems. Sanitary sewers convey domestic and industrial wastewater to the treatment facility. Storm sewers contain only stormwater and are usually discharged directly to a water body such as a river or stream. Combined sewers contain both wastewater and stormwater. Combined sewers can be problematic for some wastewater treatment facilities. After a heavy storm, the sheer volume of stormwater and wastewater combined can overwhelm the capacity of the treatment system.

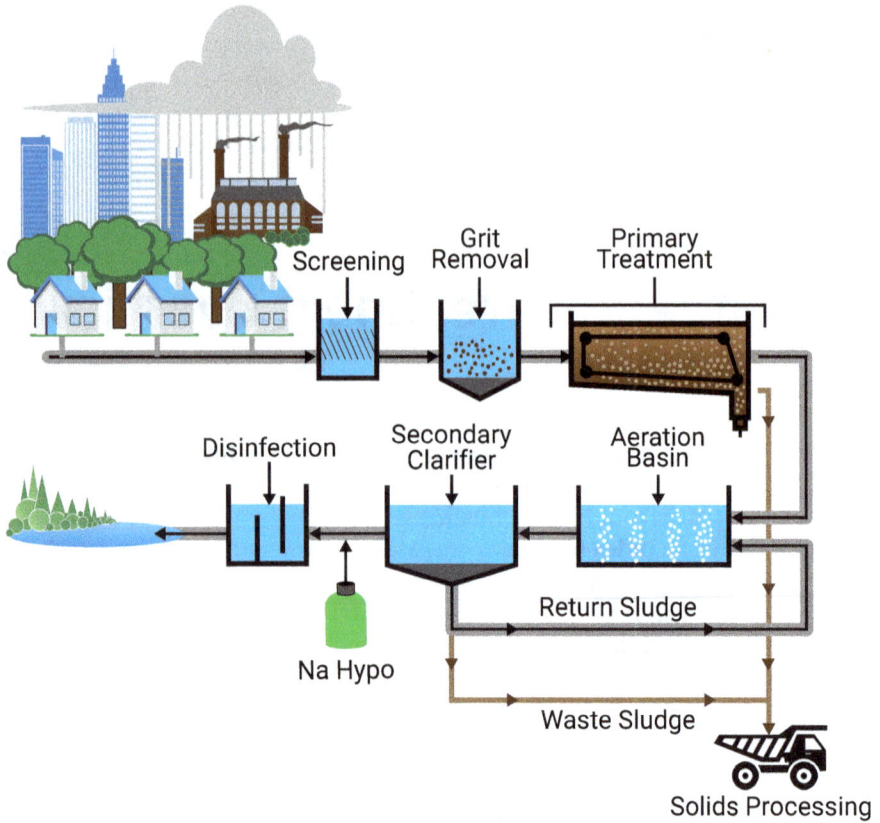

Figure 1-1   Treatment process flow

Generally, raw wastewater entering the wastewater treatment system is 99.98% water and only 0.02% solids. The wastewater flows through several stages of treatment before being discharged to the waterways (Figure 1-1) (Tchobanoglous et al. 2003).

The first stage is *preliminary treatment*. Preliminary treatment prepares the wastewater for the rest of the treatment process. It includes screening and grit removal. During the screening process, wastewater flows through bar screens to remove any large, untreatable material such as sticks, rags, cans, plastics, and other inert debris that can cause damage to pumps and interfere with the flow of water through the rest of the process. After sewage has been screened, it passes into a grit chamber, where the water is slowed down enough to allow the heavy solids like rocks, grit, and sand to settle to the bottom, while the lighter organic particles stay in suspension. A grit chamber is particularly important in communities with combined sewer systems, where sand or gravel may wash into sewers along with stormwater. All of the materials collected in screening and grit removal are removed from the facility and hauled to landfills for disposal.

The wastewater leaving the grit chamber still contains organic and inorganic particles as well as large molecules in suspension that are too small and light to settle out. In many treatment systems, the next stage is *primary sedimentation.* Although in some systems wastewater flows from preliminary treatment straight to secondary treatment, skipping primary treatment, the role of primary sedimentation is significant. Because the speed of the flow through primary sedimentation is reduced, the suspended solids will gradually sink to the bottom, where they form a mass of solids called raw primary sludge. The sludge is usually removed from the sedimentation tanks by pumping, after which it may be further treated, disposed of in a landfill, or incinerated. During the primary stage, generally a coagulant is added that aids with the settling of suspended solids. In addition, greases and oils rise to the surface and are skimmed off and removed for further treatment. In other words, the purpose of primary treatment is to settle out the "settleables" and float up the "floatables." The wastewater leaving primary treatment is called primary effluent. Primary effluent contains mostly dissolved organic and very fine organic particles. There are still facilities that discharge to the oceans that end treatment at the primary stage. However, over the years, primary treatment alone has been unable to meet many communities' demands for higher water quality. To meet the demand for higher water quality, cities and industries normally treat the wastewater to a secondary treatment level and, in some cases, also use advanced treatment to remove nutrients and other contaminants.

The next stage of treatment is called *secondary treatment.* Secondary treatment is a critical stage in wastewater treatment, focusing on the removal of biodegradable organic matter (in solution or suspension) that escapes primary treatment. This is primarily achieved through biological processes in which microorganisms metabolize the organic pollutants. The secondary stage of treatment removes about 85% of the organic matter in sewage by making use of the bacteria in it. There are several different secondary treatment technologies; however, the one thing they have in common is the use of microorganisms to remove organic waste from the water. This book will focus on the *activated sludge treatment* process.

The activated sludge process speeds up the work of the bacteria by bringing air and sludge heavily laden with bacteria into close contact with sewage. After the sewage leaves the settling tank in the primary stage, it is pumped into an aeration tank, where it is mixed with air and *"return activated sludge (RAS)"* loaded with bacteria and allowed to remain for several hours. This mixture of RAS and wastewater is termed *"mixed liquor."* During this time, the bacteria break down the organic matter into harmless byproducts. The mixed liquor flows from the aeration basin to secondary clarifiers, where time is allowed for the activated sludge (solids) to separate from the liquid. The sludge settles to the bottom, and the clarified liquid flows over the top. The sludge, now activated with additional billions of bacteria and other tiny organisms, can be used again by returning it to the aeration tank for mixing with air and new sewage, thus the term "return activated sludge."

To complete secondary treatment, effluent from the secondary clarifier is usually disinfected before being discharged into receiving water. The two most common methods of disinfection are chlorination and ultraviolet (UV) radiation. Chlorine is fed into the water to kill remaining pathogenic bacteria and to reduce odor. Done properly, chlorination will kill more than 99% of the harmful bacteria in an effluent. Many states now require the removal of excess chlorine before discharge to surface waters by a process called dechlorination. Alternatively, UV light can also be used for disinfection. The treated water flows through a channel and is exposed to UV light. When UV radiation penetrates the cell wall of an organism, it destroys the cell's ability to reproduce.

To return cleaner and more usable water to receiving streams, municipalities are increasingly implementing advanced wastewater treatment techniques, alone or in combination. Conventional activated sludge is a proven system, but determining the most effective secondary treatment system for your local wastewater treatment plant will depend on influent water quality and volume, wastewater composition, and operation and maintenance costs.

REFERENCE

Tchobanoglous, G., F.L. Burton, and H.D. Stensel. 2003 (4th ed.). *Wastewater Engineering (Treatment Disposal Reuse)*. New York: McGraw-Hill.

Chapter 2

Microbiology and Wastewater Treatment

What does microbiology have to do with wastewater treatment? Wastewater treatment is a microbiological process. Wastewater treatment systems are designed to house microorganisms and bring them into contact with bio-degradable substances in the wastewater. The microorganisms use these substances for energy, growth, and reproduction, and in exchange, these organic pollutants are removed from the wastewater. There are many differ-ent treatment technologies and just as many different configurations; how-ever, the one thing they have in common is microorganisms doing the work of removing the waste from the water. The microorganisms are the VIPs of the treatment process. The health and well-being of these microorganisms are critical to the adequate treatment of wastewater. So, it is helpful to mon-itor their activity and to understand the conditions that allow them to do their best work.

Many different types of microorganisms are present in wastewater and enter the treatment system via wastewater from homes, industries, and storm-water runoff. In the treatment system, microorganisms come in contact with the biodegradable materials in the wastewater and consume them as food. The successful removal of waste from the water depends on how efficiently the bacteria consume the organic material and on the ability of the bacteria to stick together, form floc, and settle out of the bulk fluid. The flocculation (clumping) characteristics of the microorganisms in activated sludge enable them to form solid masses large enough to settle to the bottom of the settling basin. The better the flocculation characteristics of the sludge, the better the settling and treatment efficiency.

After the aeration basin, the mixture of microorganisms and wastewater (mixed liquor) flows into a settling basin or clarifier, where the microorgan-isms that have amassed to form floc (called sludge or biosolids) are allowed to settle. Some of the sludge (containing the settled microorganisms) is continuously recirculated from the clarifier back to the aeration basin to ensure adequate amounts of microorganisms are maintained in the tank.

The microorganisms are again mixed with incoming wastewater, where they reactivate to consume and remove organic pollutants. Then the process starts again. The aeration basin is the microbiological hub of the treatment process. This is where most microbiological activity takes place. So, for the purpose of the basic activated sludge process, much of the discussion will involve microbial activity that takes place in the aeration basin.

The activated sludge process, under proper conditions, is very efficient. It removes 85–95% of the solids and reduces the organic load by about the same amount. The efficiency of this system depends on many factors, including wastewater climate and characteristics. Toxic wastes that enter the treatment system can disrupt biological activity. Wastes heavy in soaps or detergents can cause excessive frothing and thereby create aesthetic or nuisance problems. In areas where industrial and sanitary wastes are combined, industrial wastewater must often be pretreated to remove the toxic chemical components before it is discharged into the activated sludge treatment process. Nevertheless, microbiological treatment of wastewater is by far the most natural and effective process for removing waste from water.

Five major groups of microorganisms are generally found in the aeration basin of the activated sludge process.

- Bacteria
- Protozoa
- Metazoa
- Filamentous bacteria
- Algae, fungi, water fleas, and mites

Many different types of microorganisms enter the wastewater treatment system. Waste from our homes, from industry, and from stormwater and surface runoff entering the treatment system contains bacteria, archaea, protozoa, metazoa, viruses, parasites, fungi, and molds. Treatment systems are designed to "house" microorganisms and to favor the growth of those that are beneficial to the wastewater treatment process. The process is generally not favorable for the growth of archaea, fungi, molds, and most viruses. However, the process does favor the growth of bacteria, protozoa, and metazoa. We will study these microorganisms in more detail in later chapters.

Chapter 3

General Microscopy

MICROSCOPY

Most microorganisms are too small to see with the naked eye. The smallest size that can typically be seen with the human eye is 0.1 mm. Bacteria range from 0.0002 to 0.01 mm in length. In fact, if you lay 1,000 bacteria cells end to end, they will only span the tip of a pin (Postgate 2011). Protozoa are significantly larger than bacteria but are still relatively small and can range from 0.002 to 0.2 mm. Metazoa, on the other hand, are larger still. They are multicellular microorganisms and can range from 0.05 to 2.0 mm. Therefore, the activity of these microorganisms is best viewed with a microscope.

Microscopes have evolved from simple magnifying lenses to complex instruments capable of revealing the intricate details of cells, molecules, and even atoms. The early lenses in the 1st century AD were called "burning glasses" or "magnifiers." They were simple convex lenses used to magnify objects. In the 16th and 17th centuries, Hans and Zacharias Jenssen created the first compound microscope by stacking two lenses in a tube. However, these early microscopes had poor image quality. Galileo Galilei improved on the design, but a Dutch scientist known as the father of microbiology, Antonie van Leeuwenhoek, made significant improvements and built a simple, single-lens microscope with remarkable magnification up to 300 times. He was the first to observe and describe bacteria, protozoa, sperm cells, and blood cells.

In the 18th and 19th centuries, Joseph Jackson Lister developed achromatic lenses that corrected color distortion. Advances in lens manufacturing and optical theories improved image quality and led to more sophisticated and powerful microscopes. Electron microscopes and fluorescence microscopy techniques were developed in the 20th century. Developed by Ernst Ruska and Max Knoll in 1931, transmission electron microscopes use electron beams to achieve much higher magnifications (up to millions of times). The scanning electron microscope, developed by Manfred von Ardenne in 1937, provided detailed three-dimensional images by scanning samples with an electron beam. Fluorescence microscopy uses fluorescent dyes and laser technology to produce high-resolution images. Superresolution microscopes were developed in the 2000s (Science Learning Hub – Pokapū Akoranga Pūtaiao 2016). Much like high-speed jets have broken the sound barrier, these microscopes have broken the diffraction limit of light, allowing for imaging at the nanometer scale. Advancements in optics, materials science, and technology continuously expand our ability to explore and understand the once-hidden microscopic world.

While many of the advanced microscopes can observe images at the nanometer scale, these instruments are a bit too sophisticated for general wastewater microbiology. For our purposes, a simple compound microscope will do. However, even with a simple microscope, an understanding of different and slightly more advanced microscopy techniques is helpful when looking at active sludge.

A compound microscope has a set of ocular lenses and objective lenses. The ocular lens is the eyepiece. Some microscopes are monocular, with one eyepiece; some are binocular, with two eyepieces; and some are trinocular, with three eyepieces. Trinocular microscopes allow a second person to view the specimen or a camera to be attached to capture images or videos. Light from an illuminator, or light source, is passed through a condenser that directs the light rays through the microorganisms on the slide. The light rays pass through the microorganisms into the objective lens. The image is then formed on a mirror and is magnified again by the ocular lens (Figure 3-1).

Most ocular lenses (the eyepieces) magnify the specimen 10 times. There are three sets of objective lenses: low power (10× objective or less), high power (20× or 40×), and oil immersion (100×). The oil-immersion lens is used by lowering the objective and immersing it in oil before focusing it on the specimen.

How is the total magnification of a specimen determined? If you multiply the magnification of an objective with the magnification of the ocular, you can determine the total magnification of the specimen (Table 3-1). For example, for the low-power objective, the total magnification would be 10 (ocular magnification) times 10 (low-power objective) and would equal 100. For the high-power objective, the total magnification would be 400; for the oil-immersion objective, 1,000.

Under normal operation, light rays from a light source are passed through a condenser that directs the rays through the specimen. This produces a *brightfield illumination*. However, most microorganisms in wastewater are transparent. This makes distinguishing cell structures difficult because the water is also transparent. Another technique, *darkfield illumination*, is actually the opposite of brightfield illumination and makes the microorganisms appear white and the background black, much like a photograph film negative. The light does not pass through the specimen but rather hits only its sides. Brightfield illumination is best used when observing stained slides. Brightfield and darkfield illumination are the most

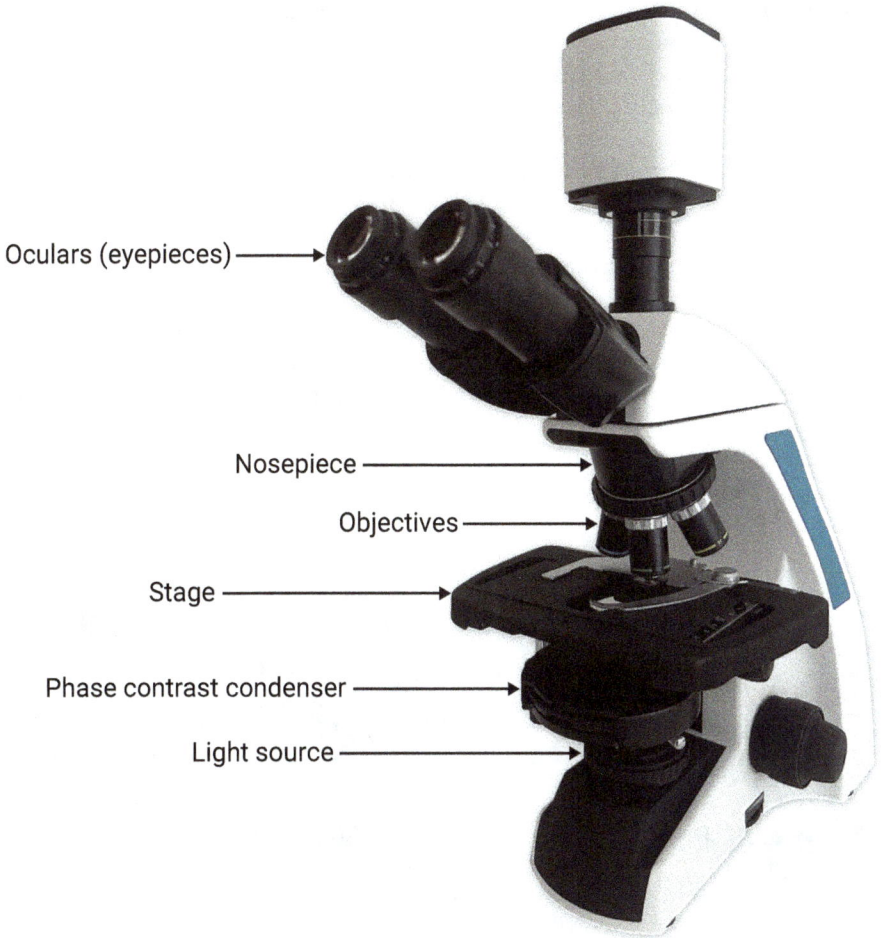

Oculars (eyepieces) ⟶

Nosepiece ⟶

Objectives ⟶

Stage ⟶

Phase contrast condenser ⟶

Light source ⟶

Figure 3-1 Typical phase contrast microscope

Table 3-1 Determining objective magnification

Objective Magnification × Ocular Magnification	Total Magnification
10× Objective × 10	100
20× Objective × 10	200
40× Objective × 10	400
100× (oil immersion) Objective × 10	1,000

basic microscopy techniques. Although microorganisms can be seen, it is difficult to see detailed structures using these techniques (Figure 3-2).

Several microscopy techniques do, however, allow clear and distinct cell structures to be seen. *Phase contrast* microscopy uses a special condenser that slows the light as it enters the denser parts of the microorganisms. This allows certain structures to stand out from other, less dense parts of the cell and from the surrounding fluid. This technique allows living organisms to be observed in more detail. Cell shape and structure are much more visible than with brightfield or darkfield illumination (Figure 3-3).

Other microscopy techniques include *differential interference contrast (DIC)*, which gives a three-dimensional appearance of the cell and other cell structures, and *electron microscopy*, which is used to examine much smaller objects such as viruses or the internal structures of cells. The magnifying power of the electron microscope is far greater than that of all other types of microscopes. Specimens as small as 0.1 nm can be observed. This is far superior to regular light microscopes, which are limited to about 200 nm.

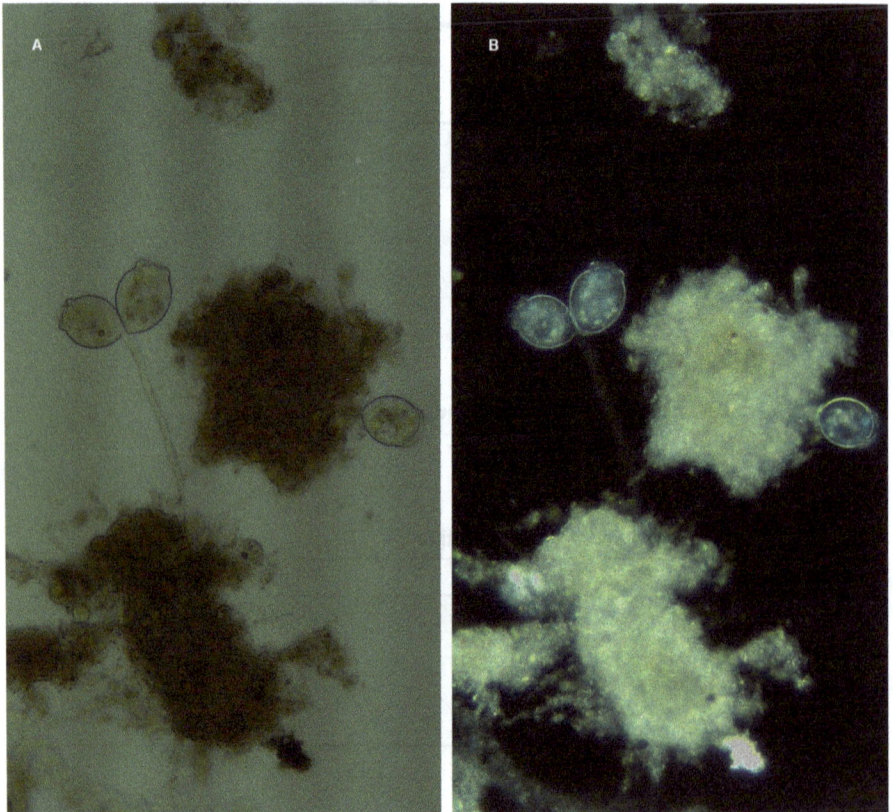

Figure 3-2 Brightfield illumination (40× magnification) (A) and darkfield illumination (40× magnification) (B)

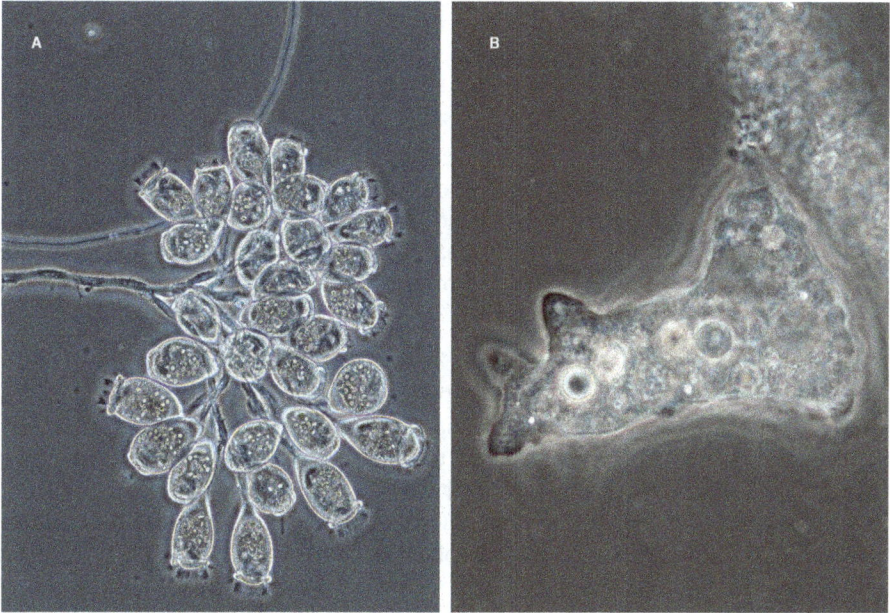

Figure 3-3   Phase contrast illumination (40× magnification) (A) and phase contrast illumination (100× magnification) (B)

Although the electron microscope is far superior and it is my dream to own one, it is extremely expensive, complicated to operate, and not necessary for general wastewater microscopy.

For observing wastewater microorganisms, brightfield and darkfield microscopes are limited in their use. While the DIC microscope provides a better image than the phase contrast microscope, it is considerably more expensive. Therefore, I recommend the phase contrast microscope for observing microorganisms in wastewater.

SAMPLE COLLECTION

When conducting microbiological analyses and observations, it is important to keep everything as consistent as possible. Samples should be collected from the same location each time. For examining activated sludge, samples should be collected from the discharge end of the aeration basin just before entering the clarifiers.

SLIDE PREPARATION AND STAINING

The laboratory techniques described in this section are for general observation techniques for operators only. To develop a laboratory program that includes tracking quantifiable changes in protozoa/metazoa populations, bacteria behavior, and treatment system conditions, procedures that include consistent volumes of sample and consistent scanning and counting procedures will be required. Those procedures can be found in my laboratory manual, titled *A Wastewater Microbiology Laboratory Manual for Operators* (2011). For general observations, the following simple procedures will suffice.

Slide Preparation

Standard slides are made of glass or plastic. For most purposes, glass slides with a thickness of 1–1.2 mm are used. When working with high-power objectives and condensers, the slide thickness should be reduced to 0.8–1 mm. A coverslip or cover glass is a very thin square piece of glass (or plastic) that is placed over the water drop on the slide. With a coverslip in place, the drop is flattened, allowing the observer to focus with high power on the specimen. The coverslip also confines the specimen to a single plane, thereby reducing the amount of focusing necessary. The coverslip also protects the objective lens from immersion into the water drop.

Wet Mount

A wet mount allows living microorganisms to be observed as they appear in the environment. Cell measurements can be taken and cell shape more accurately determined. A drop of the fluid to be examined is placed in the middle of a slide, and a coverslip is gently placed on top. It is important not to make the drop too big, which will make the coverslip float, or too small, which will quickly dry. A wet mount can be observed using any of the objectives, including the 100× objective with oil, depending on the desired magnification. It is best to use the phase contrast condenser when observing samples live. The phase contrast condenser has a setting for each objective. Each time you switch to a different objective, you should change to the appropriate phase contrast condenser setting. Usually, setting 1 is for the lower magnifications (10× or 20×), setting 2 is used with the 40× objective, and setting 3 is used with the 100× oil-immersion lens objective.

Smear Preparation and Staining

Microscopic cells, tissues, and organisms are transparent and can be difficult to see. Important information such as cell shape and motility can be observed without stains and should be done so with the wet mount. Once the sample is dried for staining, much of the structure is distorted. Staining, on the other hand, reveals different properties and provides contrast for viewing. Stains are substances that adhere to or penetrate a cell to give it color. Different stains are attracted to different types of organisms or different parts of an organism and can be used to differentiate between species or to view specific parts of organisms. To stain cells, the sample has to be fixed to the slide so that it will not wash off with water. This is done by spreading a thin film of mixed liquor or foam over the slide and allowing it to air dry. Once the sample has dried completely, it is ready for staining (Figure 3-4).

The two most common staining techniques are simple and differential. *Simple staining* uses only one color and does not differentiate between microorganisms. It just makes the microorganisms present stand out on the slide. *Differential staining* uses more than one dye and stains different types of microorganisms in different colors. Bacteria are distinguished by the way they react to the stains. Differential staining is used more often in wastewater microbiology. The most common differential stain used is the Gram stain.

Gram Stain

The Gram staining procedure was developed in 1884 by Hans Christian Gram (Sandle 2004). It is probably the most widely used procedure and helps to distinguish between two types of bacteria that otherwise would be difficult to tell apart. It divides bacteria into two large groups: (1) gram-positive bacteria that retain the color of the crystal violet dye and (2) gram-negative bacteria that lose the crystal violet color when washed with alcohol. The gram-negative microorganisms are then counterstained with safranin, which gives them a pink color. Thus, gram-positive bacteria stain purple, and gram-negative bacteria stain pink (Figure 3-5). Bacteria react differently to staining because of the difference in the structures of their cell walls. The cell wall on a gram-negative bacterium has a layer made up of lipopolysaccharide. This polymer-like substance resists the crystal violet purple dye, so the cell wall will not retain the purple color. The Gram staining procedure is done in four simple steps with four staining solutions, which can be purchased already prepared. Make sure you stain the right side of the slide (smear side up)!

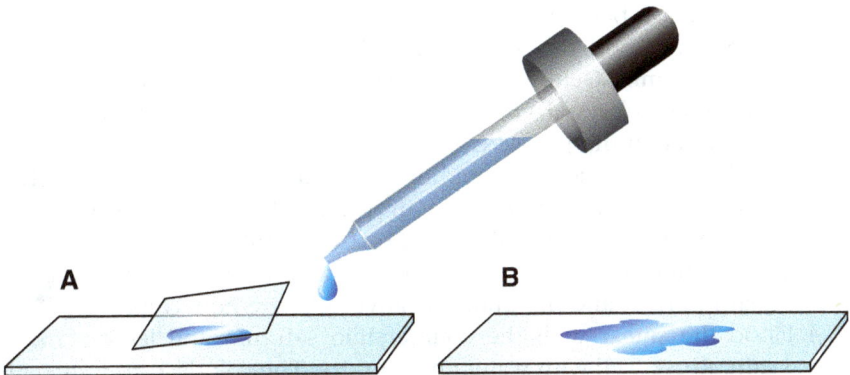

Figure 3-4 **Wet mount (place drops on slide and gently cover with a coverslip) (A) and smear (gently spread sample on slide and set aside to dry completely before staining) (B)**

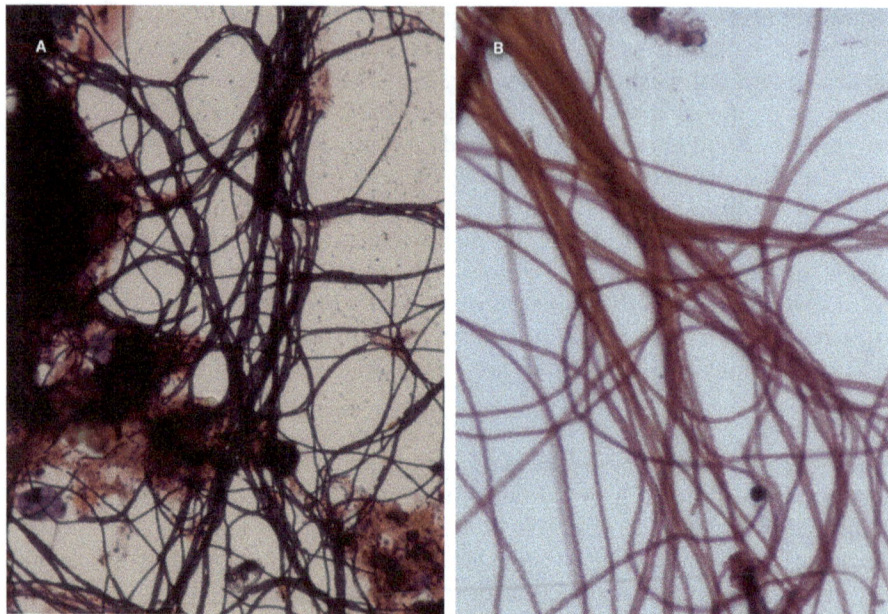

Figure 3-5   Gram-positive stains purple (A), and gram-negative stains pink (B)

The Gram stain kit contains four solutions:
- Solution 1: crystal violet
- Solution 2: Gram's iodine
- Solution 3: acetone/ethanol (50:50 volume per volume)
- Solution 4: safranin

The Gram stain procedure is as follows:
1. Completely flood the dried smear with crystal violet and let it stand for 1 min. Gently wash the slide with water. Gram-negative and gram-positive bacteria become directly stained and appear purple.
2. Completely flood the smear with Gram's iodine solution and let it stand for 1 min. Gently wash the slide with water and blot with absorbent paper. A crystal violet–iodine complex is formed that helps the bacteria retain the purple color.
3. Decolorize the smear by holding the slide at an angle and applying Gram's alcohol solution drop by drop until the violet color washes off the slide. This generally takes 15–25 s. Be careful not to over-decolorize. Quickly blot the slide with absorbent paper. The gram-positive cells remain purple, and the gram-negative cells are now colorless.
4. Flood the smear with the counterstain safranin and let it stand for 1 min. Rinse well with water and blot dry. The colorless gram-negative bacteria are stained pink by the safranin.

5. Examine under oil immersion using the 100× objective with bright-field illumination (do not use phase contrast).

Neisser Stain

Another differential stain commonly used with activated sludge samples is the Neisser stain (Ligon 2005). This stain was named after Albert Neisser, "affectionately" called the father of gonococcus, the bacterium responsible for gonorrhea. He was also the first to stain and demonstrate leprosy bacteria. The Neisser stain also divides bacteria into two large groups: those that store polyphosphate granules and those that do not. This stain is most commonly used to identify phosphorus-accumulating organisms in biological phosphorus removal systems. Neisser-positive bacteria stain blue; Neisser-negative bacteria stain brown (Figure 3-6).

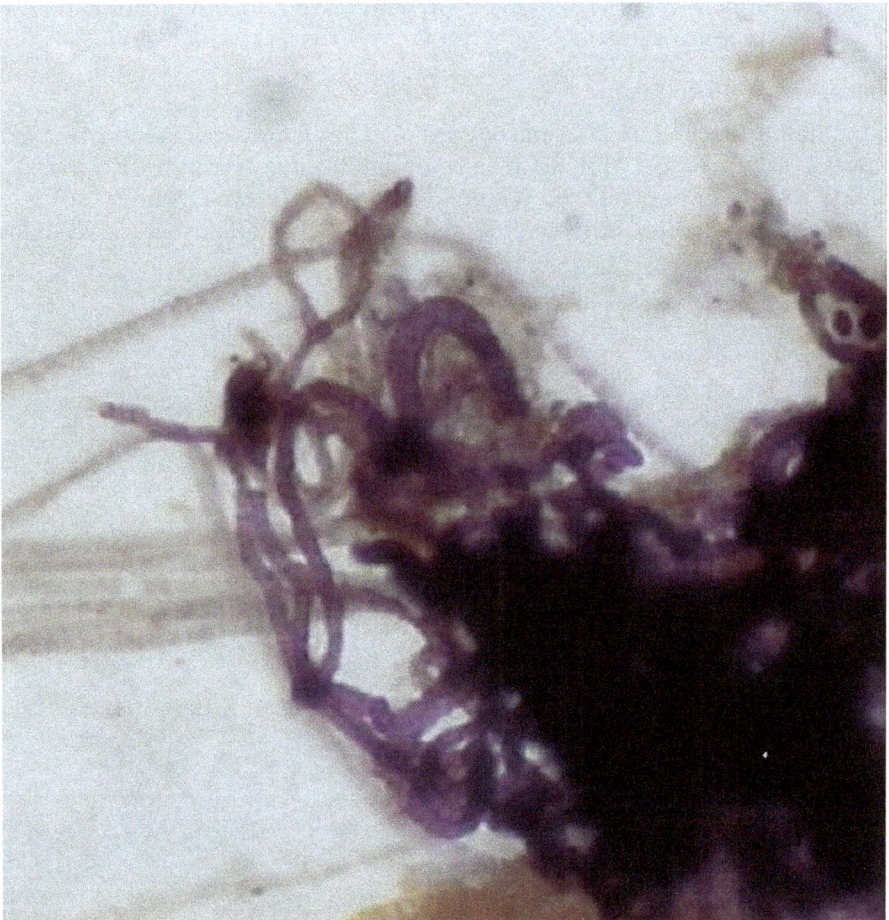

Neisser-positive stains purple, and Neisser-negative stains brown.

Figure 3-6 Neisser-positive and Neisser-negative bacteria

Neisser-positive granules can also be seen within the cells of some bacteria. Neisser staining solutions can be purchased in a kit already prepared. The Neisser staining kit contains three solutions:

- Solution 1 A: methylene blue
- Solution 1 B: crystal violet
- Solution 2: Bismark brown

The Neisser stain procedure is as follows:

1. First, you must prepare solution 1 by mixing two parts by volume of methylene blue with one part by volume of crystal violet. This solution must be prepared fresh monthly.
2. Cover a smear that has been thoroughly air dried with solution 1 for 30 s and rinse with water for 1–2 s.
3. Cover the smear with Bismark brown for 1 min, rinse well, and blot dry.
4. Examine with 100× oil immersion with brightfield illumination (do not use phase contrast).

Sudan Black (PHB) Stain

The Sudan black stain was introduced as a fat stain by L. Liston in 1934 (Booth et al. 2000). This stain is a lipophilic dye, meaning it binds to lipids within and on the cell (Figure 3-7). Polyhydroxybutyrate (PHB) is a biodegradable lipid-like polymer stored by some bacteria as an intracellular

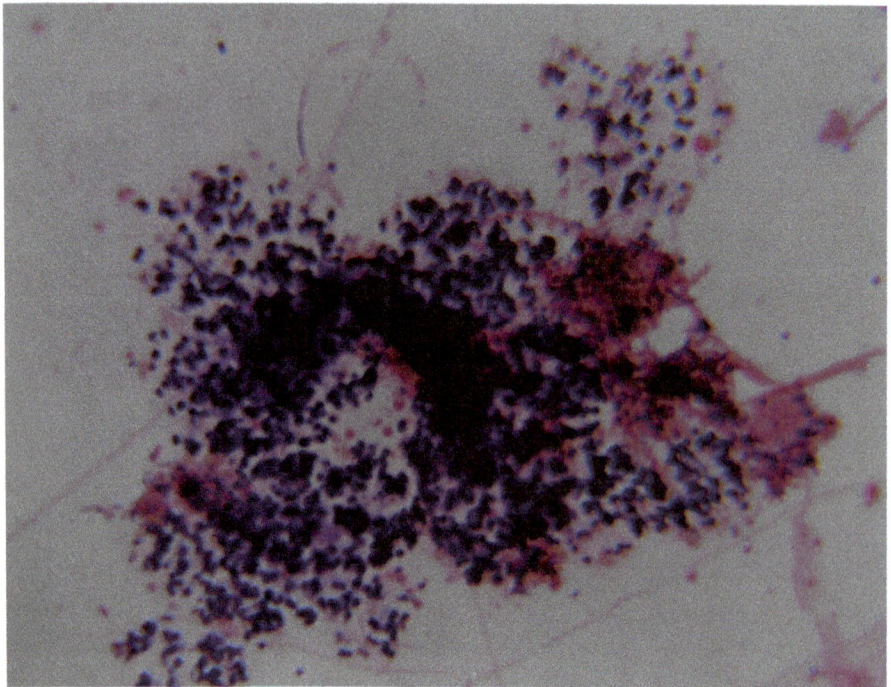

Figure 3-7　Intracellular PHB within the cell

reserve for carbon and energy. The ability of Sudan black to dissolve in fatty material allows it to detect PHB granules, which would otherwise be invisible under light microscopy. Sudan black (PHB stain) staining solutions can also be purchased in a kit already prepared.

The Sudan black staining kit contains two solutions:
• Solution 1: Sudan black
• Solution 2: safranin O

The Sudan black stain procedure is as follows:
1. Cover a smear that has been thoroughly air dried with Sudan black for 10 min and rinse with water for 1–2 s.
2. Cover the smear with safranin O for 10 s; then, rinse well and blot dry.
3. Examine with 100× oil immersion with brightfield illumination (do not use phase contrast).

India Ink Stain

India ink is used to identify the presence of exocellular lipopolysaccharide (slime) that is produced by bacteria under low-nutrient or extremely low-food conditions. The carbon black particles in the India ink cannot penetrate the slime, so when viewed under the microscope, the background fluid appears black, and the slime appears white (Figure 3-8).

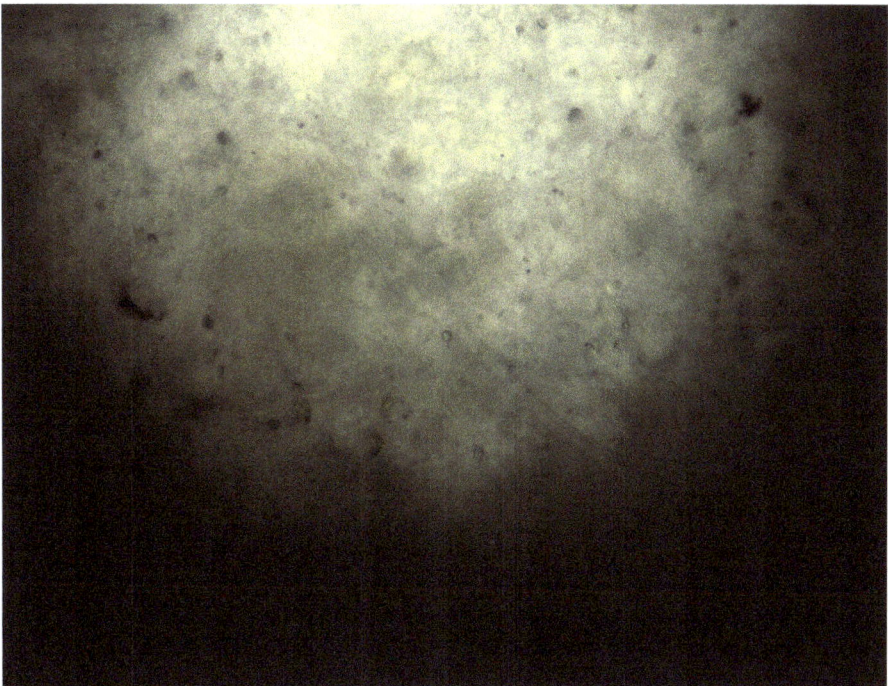

Figure 3-8   Positive India Ink stain

The India ink stain procedure is as follows:

1. Place one or two drops of mixed liquor on a clean slide.
2. Add one drop of India ink and cover gently with a coverslip.
3. Gently blot to remove excess fluid so that the coverslip will be flat against the slide.
4. Slide can be viewed using the 40× objective or 100× oil-immersion objective.
5. Observe using brightfield using maximum light (do not use phase contrast).

USING THE MICROSCOPE

1. Rotate the revolving nosepiece and place the low-power objective above the slide on the microscope stage. Adjust the objective until it is about 1 in. above the stage.
2. Place the slide on the stage. Make sure the smear or wet mount is facing up and centered directly under the objective.
3. Peering through the eyepiece, turn the coarse adjustment knob and move the objective closer to the slide until the specimen begins to come into focus.
4. Turn the fine adjustment knob to bring the specimen into sharp focus.
5. Once the specimen is focused with the low-power objective, rotate the nosepiece to place the higher-power objective over the specimen. The specimen should be nearly in focus. Adjust the focus with the fine adjustment knob.
6. To move to the oil-immersion lens, rotate the nosepiece until it is between objectives. Place a drop of oil directly on the coverslip on the wet mount (or directly on top of the stained smear). Slowly rotate the 100× objective into the oil. The image should be nearly in focus. Bring the image into sharp focus with the fine adjustment knob.

Procedures for observing microorganisms with a microscope can range from fairly simple to very complex. The average operator cannot afford to spend hours observing microorganisms under the microscope. The attempt here is to keep it simple.

REFERENCES

Booth, G., H. Zollinger, K. McLaren, W.G. Sharples, and A. Westwell. 2000. "Dyes, General Survey." In *Ullmann's Encyclopedia of Industrial Chemistry*. Weinheim, Germany: Wiley-VCH. https://doi.org/10.1002/14356007. a09_073

Ligon, B.L. (2005). "Albert Ludwig Sigesmund Neisser: Discoverer of the Cause of Gonorrhea." *Seminars in Pediatric Infectious Diseases*. 16(4): 336-341. doi:10.1053/j.spid.2005.07.001

Postgate, J. 2000. *Microbes and Man.* Cambridge, Mass.: Cambridge University Press.

Sandle, T. 2004. "Gram's Stain: History and Explanation of the Fundamental Technique of Determinative Bacteriology." *IST Science and Technology Journal.* 54:3-4.

Science Learning Hub – Pokapū Akoranga Pūtaiao. 2016. "History of Microscopy – Timeline." www.sciencelearn.org.nz/resources/1692-history-of-microscopy-timeline (accessed April 26, 2025).

Chapter 4

Bacteria

Bacteria are amazingly complex, although they are single-celled microorganisms. Some bacteria can even grow in extreme cold or extreme heat—temperatures beyond which any human can survive. Bacteria are among the most abundant organisms on Earth and can be found on our skin, in our mouths, in our intestines, in our food and drinks, in the air, in water, and in soil. They can be found in the highest altitudes, in the deepest oceans, in the Arctic, and in volcanoes. We often associate bacteria with disease, but most are harmless, and many provide valuable services for humans. Bacteria play essential roles in maintaining our health. Gut bacteria help digest complex carbohydrates, fibers, and other nutrients that our bodies can't break down on their own and also help prevent the growth of harmful disease-causing bacteria (pathogens). Certain bacteria produce essential nutrients like vitamin K and some B vitamins. Others occupy niches in our body (e.g., skin, gut, respiratory tract) to prevent harmful microorganisms from taking hold. Bacteria on our skin help maintain the skin's protective barrier and help keep harmful microbes in check (so don't overdo it with the hand sanitizers). In addition, bacteria in the soil, water, and sewage systems transform human waste into reusable materials.

Bacteria are classified based on various characteristics, including metabolism, shape, and their oxygen (O_2) requirements. They exhibit two main metabolic pathways: autotrophic and heterotrophic. Autotrophs produce their own food using light energy (photoautotrophs) or chemical energy (chemoautotrophs). Heterotrophs obtain energy by consuming organic matter. Most of the bacteria in the activated sludge process are heterotrophs. Treatment systems that are required to biologically remove ammonia will also contain significant numbers of autotrophs (nitrifying bacteria).

Bacteria can also be classified based on how they respond to oxygen. *Aerobic* bacteria require free oxygen for growth and maintenance. *Facultative* bacteria can grow in the presence or absence of free oxygen, and *anaerobic* bacteria do not require oxygen for growth and may even be harmed or killed when oxygen is present. The most important microorganisms in the traditional activated sludge process are aerobic bacteria. However, in treatment systems designed with anaerobic or anoxic zones for biological phosphorus

(P) removal, anaerobic and facultative bacteria play an essential role (more discussions on this later).

Bacteria make up about 95% of the microorganisms in activated sludge. They are introduced into the treatment system via surface runoff, wash water, and feces. Because they are primarily responsible for removing organic compounds from the wastewater, it is important to understand the conditions that make them work their best. Let's begin by understanding how they are structured.

CELL STRUCTURE

Bacteria are small, relatively simple, one-celled microorganisms. They generally have one of three basic shapes: coccus (spherical or oval shaped), bacillus (rectangular, rodlike, or discoid), and spirillum (spiral or corkscrew shaped) (Figure 4-1).

Some are individual cells; others form clusters, pairs, chains, squares, or other groupings. Coccus-shaped bacteria (the word *coccus* means berries) occur singly or in pairs called diplococci, attached in short chains called streptococci, in groups of four called tetrads, in three-dimensional cubes of eight called sarcinae, or in grapelike clusters called staphylococci (Figure 4-2). Most bacillus-shaped bacteria (the word *bacillus* means little

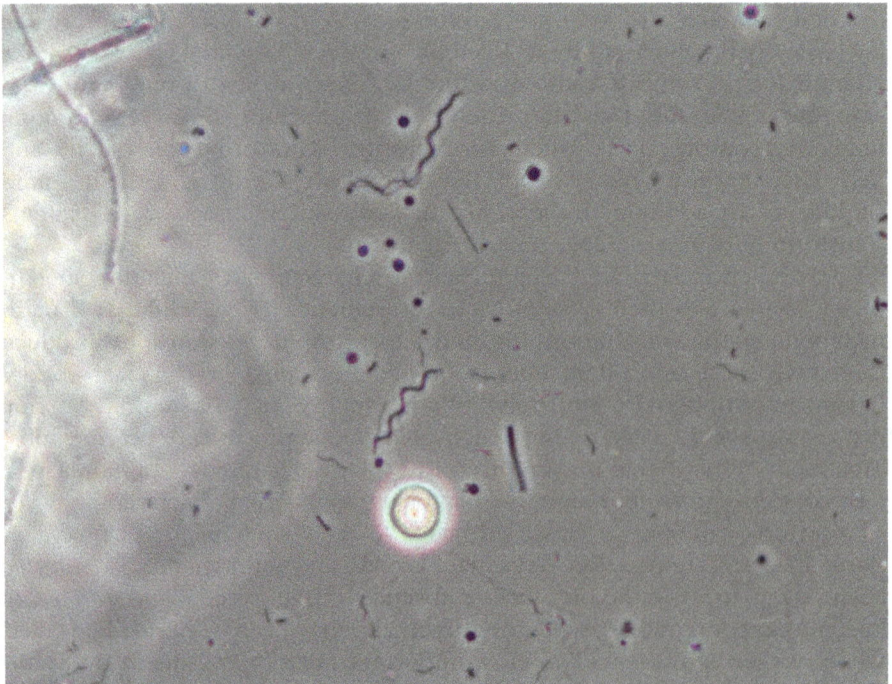

Figure 4-1 Bacteria of different shapes (100× magnification; oil immersion; phase contrast)

staff) appear as single rods. They can grow in pairs called diplobacilli or short chains called streptobacilli. They also occur in short, almost oval rods called coccobacilli (Figure 4-3). Spiral-shaped bacteria are never straight. Some, called vibrio, are shaped like little commas; spirilla have rigid bodies and are shaped like corkscrews. Spirochetes are also corkscrew shaped but with more flexible bodies (Figure 4-4).

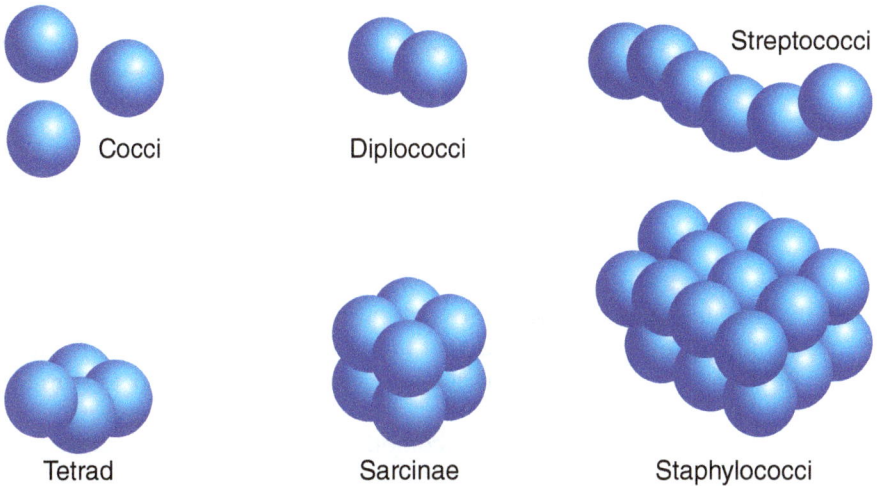

Figure 4-2 Cocci-shaped bacteria can grow as single round cells, in groupings of two cells (diplococci), as a string of cells (streptococci), arranged in groups of four cells (tetrads), groups of eight cells (sarcinae), or in larger groups of cells (staphylococci).

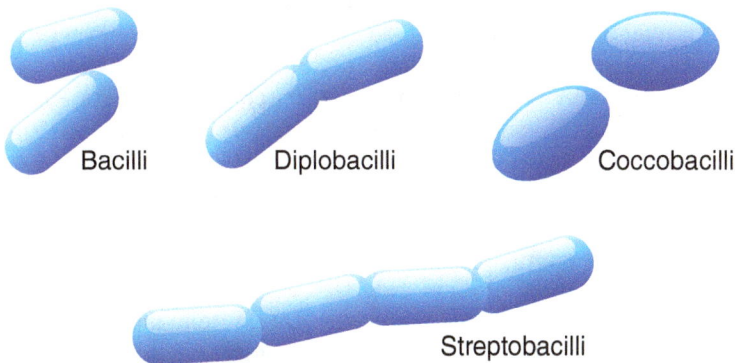

Figure 4-3 Bacillus-shaped bacteria can appear as single rod-shaped cells, in groupings of two cells (diplobacilli), as rounded rods (coccobacilli), or arranged in a string of rod-shaped cells (streptobacilli).

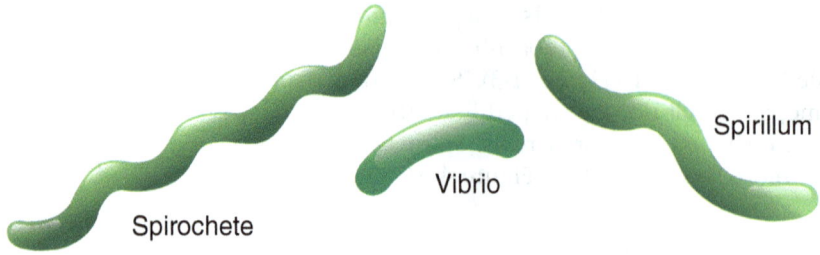

Figure 4-4 Spiral-shaped bacteria can appear as comma-shaped cells (vibrio), rigid corkscrews (spirillum), or flexible corkscrew-shaped cells (spirochetes).

A typical bacterial cell is surrounded by a semirigid cell wall that gives it its shape and protects its internal parts. The composition of the cell wall varies and can be used to classify bacteria into two major groups: gram-positive and gram-negative. Just inside the cell wall is the cell membrane. This controls the movement of substances in and out of the cell and functions as an entrance and an exit for materials that enter and leave the cell. This membrane is selective, which means that some substances cannot pass through (Figure 4-5).

Bacteria have several means of locomotion. Some may wiggle or squirm; some glide through their media; others propel themselves using flagella. Flagella are long, whiplike structures that enable bacteria to move through the water. Flagellated bacteria can move in one direction, they can tumble, and they can move in reverse. Most bacteria are structured to function very well in the activated sludge environment. They are nature's ideal employees. They can use most of the organic material commonly found in sewage. They are designed to swim to where the organic nutrients are, they can multiply to meet the demand, and they know how to settle out only after the job is done.

Most bacteria in the activated sludge consume organic compounds for energy, cell maintenance, and reproduction. Sewage from homes, industries, and stormwater that enter the wastewater treatment system contains many types of organic substances. Some bacteria, however, use inorganic compounds such as iron, sulfur, and ammonia. Bacteria require essential elements such as carbon, nitrogen, phosphorus, sulfur, and water to build cell components. Other substances such as carbohydrates, sugars, and fats are used mostly for energy. Bacteria also need minerals such as magnesium, calcium, potassium, iron, and chloride.

Some substances entering the treatment system are in the form of complex molecules, while others are simple soluble (dissolved) compounds. Only soluble compounds can pass through the bacteria cell membrane, so dissolved organic material with smaller simple molecules passes easily through the cell membrane. Larger, undissolved compounds have to be hydrolyzed (broken down) into much smaller units before they can be transported into

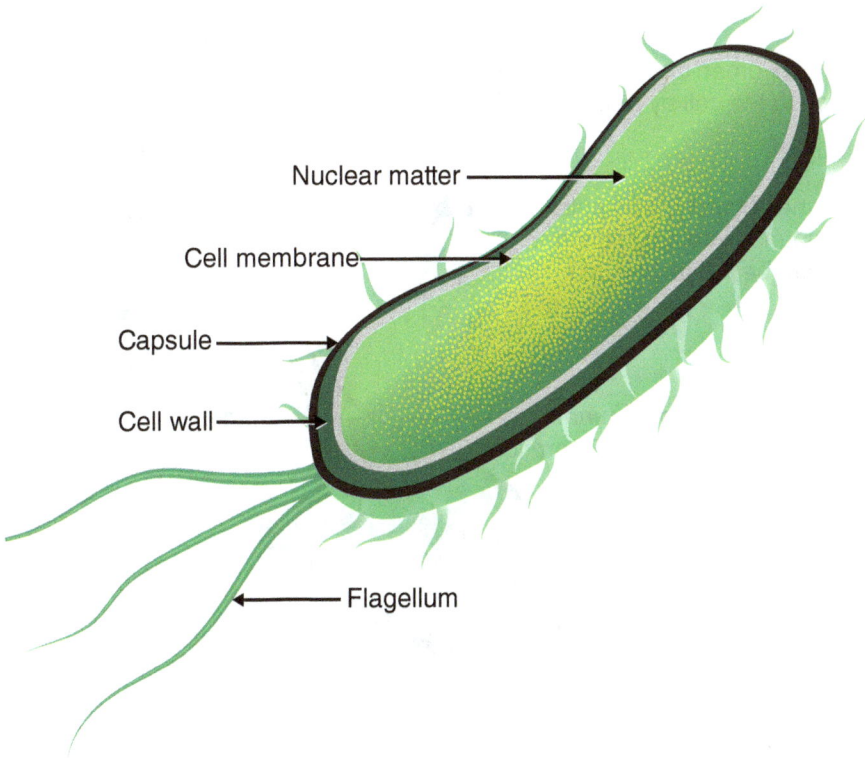

Figure 4-5 **A typical bacterial cell contains nuclear material and is surrounded by a selective cell membrane that allows substances to enter and exit the cell. The rigid cell wall gives the cell it's shape, and the flagellum is used for locomotion.**

the cell. In this case, larger particles adhere to the sticky slime layer of the cell wall because they cannot enter the cell membrane. This process is called *adsorption*. Next, the cell membrane secretes enzymes that cause chemical reactions that break the large particles down into smaller units that can be transported across the cell membrane and into the cell. This is the process of *absorption* (Figure 4-6). These complex organic substances cannot enter the cell membrane without enzymes. Many species of bacteria produce enzymes capable of converting insoluble or complex organic compounds to soluble forms. So, just because bacteria come in contact with organic substances, it doesn't mean they can consume them right away. While simple, soluble compounds are used right away; the more complex and insoluble the composition of the wastewater entering the system is, the more time it will take for the bacteria to consume it. Imagine eating pecans. We cannot eat them if they are still in the shell. We have to crack open the shell to eat the actual nut meat. It's the same with bacteria—the soluble compounds are

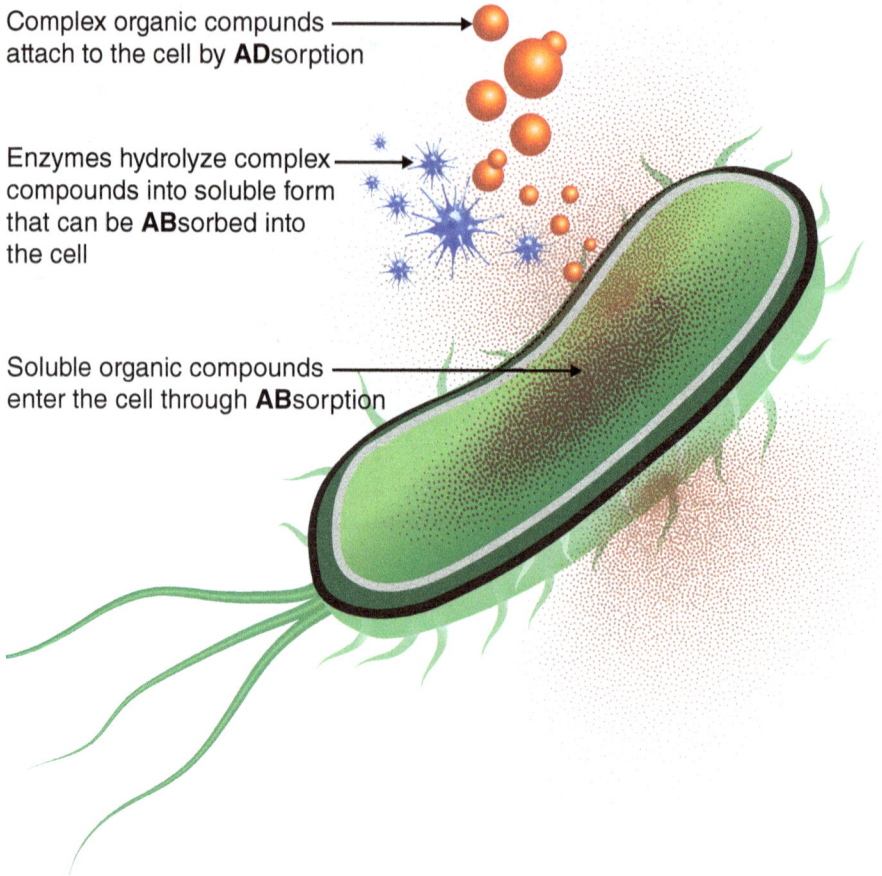

Complex organic compunds attach to the cell by **AD**sorption

Enzymes hydrolyze complex compounds into soluble form that can be **AB**sorbed into the cell

Soluble organic compounds enter the cell through **AB**sorption

Figure 4-6 Complex molecules are adsorbed to the cell wall where they are hydrolyzed into soluble form by enzymes. The soluble substances are easily absorbed into the bacteria cell wall.

like nut meat. They can consume it right away. However, insoluble complex organic compounds are like the pecan in the shell. They have to crack it open (hydrolyze) before they can consume it. Bacteria produce enzymes to hydrolyze or help break down complex organic compounds.

ENZYMES

Like all living organisms, bacteria carry out a process called metabolism, the sum of all the biochemical reactions that occur in a living cell. These reactions are necessary for life, growth, reproduction, and even death. Enzymes are proteins that trigger biochemical reactions in living organisms. All living organisms can produce thousands of enzymes. Some, called exoenzymes, perform their tasks outside the cell wall; others, called endoenzymes, perform their tasks inside the cell. Enzymes are protein and life process chemicals that can be affected by heat, acid, and alcohol and destroyed by toxic substances. Enzymes are very efficient. Under favorable conditions, they can increase the rate of biochemical reactions 10,000,000,000 times more rapidly than reactions without enzymes. Enzymes are catalysts, which means they help biochemical reactions to occur more rapidly without being consumed in the process. Thousands of enzymes in one cell must function properly, or the cell will stop functioning. Enzymes also help determine the specific traits of each organism. They distinguish the aerobic bacteria from the anaerobic bacteria; they distinguish those that prefer acidic environments from those that prefer basic environments and those that prefer heat from those that prefer midrange temperatures. Without enzymes, microorganisms cannot break down nutrients or perform cellular processes.

Many types of bacteria have the ability to produce enzymes. These are called hydrolytic bacteria. When they come in contact with wastewater, they go to work, producing the enzymes needed to break down specific complex substances. For example, they produce the enzyme proteinase to help break down proteins in the wastewater. They produce lipase to help break down fats or glucosidase to help break down complex carbohydrates. The function of enzymes may be altered by changes in the environment in the wastewater treatment system. Environmental factors such as the concentrations of enzymes, nutrients, temperature, pH, and the presence of chemical substances that serve as enzyme inhibitors all affect enzyme function. Before running out to buy a bag of enzymes, the best course of action is to determine what may be affecting enzyme function in your treatment system first.

Commercial Enzymes

Enzymes are commercially available for use in the wastewater treatment system. Commercial enzymes can be useful when the wastewater is rich in a certain substance from a particular source. Enzymes are substance specific, so for example, the enzyme proteinase can be used to help break down protein-rich waste from meatpacking industries. The enzyme lipase can be used to help with wastes rich in grease and fats. In this situation, enzymes should be added before the aeration basin so they can break the protein-rich compounds into smaller units that the bacteria can easily consume. The higher the concentration of enzymes, the faster the breakdown of the substrate.

Superbugs

Superbugs are selectively or genetically engineered to do one thing better than normal bacteria. They can produce extra enzymes that break down certain substrates. Genetically altered bacteria can significantly expedite the breakdown of certain substances. For example, genetically altered bacteria are available for use in oil spills. Most municipal systems do not need enzyme supplements or superbugs under normal operating conditions. No superbug or enzyme alone can break down all the substances found in wastewater. They can be useful, however, when dealing with extremely difficult-to-treat waste products.

As discussed, many different species of bacteria enter the treatment system. However, the types that are most important for the adequate treatment of wastewater are the *floc-forming bacteria*.

Floc-Forming Bacteria

The key to good treatment in the activated sludge system is the separation of the biological solids from the liquid portion of the treated wastewater. To get good separation, the development of a healthy biological floc is required. To develop a healthy biological floc, a healthy population of floc-forming bacteria is required. Floc formers are aerobic bacteria that, under the right conditions, can accumulate an outer "slime layer" that enables them to amass to form biological solids (floc) heavy enough to settle. They play an important role in making sure there is a good separation between the solids and liquid portion of the wastewater.

UNDERSTANDING FLOC FORMATION

Floc formers and other heterotrophic bacteria make up 95% of the microorganisms in the process. Bacteria are responsible for removing most of the organic compounds from wastewater and are therefore the most important microorganisms in the treatment system. They consume soluble organic compounds and use them for energy, cell maintenance, and the production of new cells. Floc formation occurs naturally with increasing sludge age. Floc formers are able to produce extracellular polymers (slime layer) that enable them to "stick" together or agglomerate. It is important to note that there are many species of bacteria present in the treatment system that do not have the ability to form floc.

When floc-forming bacteria come into contact with organic compounds (food), they go through five growth phases as time passes and the sludge ages: lag phase, accelerated growth phase, declining growth phase, stationary phase, and death phase.

- *Lag phase*: The lag phase is the adaptation period. Bacteria are adjusting to their new environment. The length of this phase can vary based on how different the conditions are from the conditions the bacteria came from. When bacteria are transferred from one environment to another, they have to adjust to changes in temperature, dissolved oxygen, pH, etc. They also need to produce the enzymes needed to break down the organic substrate in the wastewater. If the environment is the same, the lag phase will be shorter (not much adjustment is needed). If the environment is very different, the lag phase will take longer. During this phase, there is no growth or reproduction. It is important to understand that during this phase, bacteria are not consuming organic substrate (food) at all. In other words, they are not removing biochemical oxygen demand (BOD).

It is difficult for the bacteria to become acclimated in a treatment system that has variable loads. They need time to develop the proper enzymes. If the loading changes drastically from moment to moment, the bacteria cannot adjust. Actually, the lag time is extended. This means that there is a longer period before the bacteria can begin to consume the organic material from the water. Ideally, in such cases, an equalization tank or even primary clarification would help stabilize the loading.

- *Accelerated (exponential or log) growth phase*: Once the bacteria have accumulated all they need for growth, they begin to divide exponentially. One cell becomes two cells, become four, become eight, etc. During this phase, bacteria begin to consume organic compounds. They divide constantly, and the population doubles exponentially. The cells are active, and during this phase, food is reduced at the fastest rate in the entire process. During this phase, bacteria are motile and dispersed and do not form floc. If the treatment system were allowed to remain at this phase, the mixed liquor solids would settle poorly in the clarifier.
- *Declining growth phase*: Because of the accelerated growth, during this phase, most of the organic material has been removed. Then, because there is much less food and many more bacteria, the growth rate begins to slow down.
- *Stationary phase*: As bacteria continue to multiply, at some point the number of bacteria will begin to exceed the available food, and there will no longer be enough essential nutrients for their growth. With little food left, the bacteria can no longer increase in number. The numbers are relatively constant. The number of dying cells is about equal to the number of reproducing cells. So, the bacteria enter the stationary phase.

This is the critical stage for activated sludge treatment. It is during this phase that floc formers take steps to preserve energy and adapt to their new starvation conditions. First, they shed the flagella that enabled them to swim through the fluid. Next, they begin to produce and surround themselves with extracellular components that enable them to stick together to form floc.
- *Death phase*: During this phase, the number of viable (living) cells decreases rapidly. Cells begin to lyse (break open) and release their cellular contents. Very little treatment occurs during this stage.

If the detention time in the aeration basin is too long and allows most bacteria cells to enter the death phase, the once-healthy floc solids will begin to break apart. The solids will not settle well in the clarifier, and this will produce an unacceptably turbid effluent.

GROWTH IN ACTIVATED SLUDGE

As long as enough nutrients are available, bacteria can multiply very rapidly by splitting into two identical cells. Some can divide in two in only 11 min. Many can double in 20–30 min. In fact, if enough nutrients were available, *Escherichia coli* bacteria, under ideal conditions, could produce a mass of bacteria greater than the total mass of Earth in three days (Bruslind 2017). Bacteria are much like us: If lots of nutrients are available, they use them mostly for growth and multiplication. Although we do not split in two, we do increase in bulk. Bacteria, on the other hand, will simply continue to divide and increase in number.

Let's consider the flow of mixed liquor through the aeration basin. Wastewater is mixed with the microorganisms in the return sludge at the beginning of the aeration basin, where the level of food is quite high. Because an abundance of food is available, bacteria are growing, motile, and extremely active. Most have flagella, which enable them to swim to find available food. At this point, the bacteria are multiplying rapidly. Essential to good treatment is the formation of sludge floc. Actively multiplying bacteria cannot form sludge floc. However, as the level of food decreases, the bacteria begin to conserve energy, and they lose the flagella. Because little food is available, they do not waste energy trying to swim around to find it. Next, they begin to form a thick outer slime layer. As they bump into each other, they stick together. At first, they form small clumps, then masses large enough to settle.

Bacteria must remain in the aeration basin long enough to remove most of the food so floc can form. The amount of time that bacteria reside in the basin is called the solids detention time. Detention time can also be defined as the amount of time it takes for the water to flow from one end of the basin to the other. This is the hydraulic detention time. If the detention time is not long enough, the amount of food remaining will still be high, the bacteria will actively move and reproduce, and the discharge from the aeration basin into the settling basin will be full of free-swimming, dispersed bacteria. The result will be poor-settling sludge and a turbid effluent. If, however, the detention time in the aeration basin is long enough to allow the bacteria to consume most of the food, they will lose flagella, form a sticky slime layer, clump together, and form floc.

Sludge age is critical to floc formation. Enough time must be allowed in the aeration basin for bacteria to consume most of the organic material. The shedding of flagella and the production of the extracellular components (slime layer) that enables them to stick together to form floc only occur when bacteria take steps to conserve energy (reach the stationary phase). They will only take those steps if food is limited, and that will only occur when enough time is allowed in the basin.

BACTERIA IN WASTEWATER TREATMENT

Among all the types of microorganisms involved in the wastewater treatment plant, bacteria are the most numerous and the most important. Most bacteria in the activated sludge process that form floc are gram-negative and rod shaped. The operator's primary role is to maintain conditions in the treatment system that favor the growth of the bacteria that are best suited to remove the organic material from the wastewater. An operator can control some conditions that affect their growth, but not all. The operator has very little control over the influent characteristics of the wastewater. Influent characteristics such as temperature, nutrients, pH, toxicity, and even surfactants can affect the microorganisms in the treatment system.

Temperature

Temperature is one of the most important factors affecting bacterial growth and survival. Bacteria can grow at temperatures from below freezing to more than 100°C. Based on the temperature best suited for their growth, bacteria are classified as *mesophiles* (20–40°C), *psychrophiles* (0–20°C), *thermophiles* (40–90°C), or *extreme thermophiles* (greater than 80°C). Bacteria most suited to the environment of the wastewater treatment system are mesophiles. Mesophiles have an optimal temperature of around 37°C.

Bacteria have well-defined temperature ranges. They have a minimum, optimal, and maximum range. At minimum temperatures, cell metabolism is mostly inactive. At the optimal temperature, bacteria will grow at their fastest rate. At the maximum temperature, cells will not grow at all. For example, although temperature may be above freezing, if it is lower than the minimum temperature range for a particular bacterium species, cell growth will not occur.

While most bacteria grow best within their own species' defined temperature ranges, in the real world, bacteria can expect frequent shifts in temperature. Although the temperature in the treatment system can be relatively stable, changing gradually over the seasons, drastic changes can still occur.

The enzyme activity of the cell depends on the temperature of the environment. If the enzymes in the cells are not functioning properly, the bacteria will not function properly in the treatment system. Like most chemical reactions, the rate of a reaction that is catalyzed by enzymes increases as the temperature increases. A 10°C rise in temperature will increase the activity of most enzymes by 50–100%. A change in temperature as small as 1 or 2°C can cause a 10–20% change in results. It is important to remember that although a rise in temperature increases the rate of a reaction, enzymes can still be adversely affected by temperatures that are too high (in most cases, greater than 40°C).

Therefore, temperature directly affects the rate at which biochemical reactions occur. Biochemical reactions occur much faster in warmer temperatures. Therefore, fewer microorganisms are required to remove nutrients when the temperatures are warmer. Conversely, in colder temperatures, more microorganisms are required to do the same amount of work as is required in warmer temperatures. Most microorganisms associated with sewage perform best at temperatures in the 20°C range.

Often, as operators, we follow certain "rules of thumb." We know microorganisms work slower during colder months, so we routinely increase our "bugs" (mixed liquor concentration) during the winter. This strategy is great if the organic load is consistent from summer to winter. However, if the load decreases during the winter, then there will be no need to increase the bugs.

Nutrients

Bacteria require essential elements and nutrients to grow, build cell components, and maintain cell functions. They need carbon, nitrogen, phosphorus, and sulfur in addition to various metal ions such as magnesium, calcium, iron, and copper. Many of these nutrients are essential for enzyme reactions in the cell. Generally, domestic wastewater contains enough carbohydrates, proteins, and fats, which supply most of the nutrients that bacteria require. However, wastewater from most industries lacks many essential nutrients for proper cell growth and multiplication. When critical nutrients are absent from wastewater, the bacteria will not perform well. We will discuss the effects of nutrient deficiency later in the book. Operators have very little control over the discharges from homes and even stormwater runoff that enters the treatment system. The operator can, however, add nutrient supplements when essential nutrients like phosphorus or nitrogen are lacking.

pH

Bacteria exhibit various tolerances to pH. All bacteria require a certain pH range, which can vary widely in the many different types of bacteria. Whether bacteria can live outside of their optimal pH range depends on the particular bacteria's ability to "self-correct" during environmental pH changes. Most bacteria are neutrophiles (grow best at a pH range of 6–8). Some bacteria are *acidophilic* or acid tolerant (prefer low pH). These cannot function under neutral conditions. *Alkaliphiles* are alkaline tolerant (prefer high pH). In general, most bacteria can tolerate a high pH better than they can tolerate a low pH.

Extreme changes in pH will affect bacterial growth. Changes in pH levels have the most noticeable effects on bacterial enzymes. For each enzyme, there is an optimal pH. The optimal pH value will vary greatly from one enzyme to another. Extremely high or low pH values generally result in complete loss of activity of most enzymes. Fluctuations or rapid changes in pH will alter or denature the structure of the enzymes. Filamentous bacteria, fungi, and yeast can compete better than floc-forming bacteria in low-pH conditions.

Toxicity

Incoming wastewater carrying toxic substances can also affect the microorganisms in the treatment process. Some toxic substances may kill the microorganisms; others may retard their ability to remove nutrients from the wastewater. For example, the toxic effects from petrochemical poisoning will cause bacterial cells to develop a "protective wall." The wall prevents the cells from sticking together to form floc. Thus, they do not settle well in the clarifier. Once bacteria in the floc are dead, the floc will break apart. Dead bacteria do not form into floc. Toxic substances generally have a greater effect on higher life forms, like stalked ciliates and rotifers. They may have the advantage when competing for food because they have the ability to draw in large amounts of wastewater. However, for the same reason, they are at a disadvantage in the presence of toxic substances.

Surfactants

Surfactants, which are substances that reduce the tension on the surface of the water, make water slippery. Although surfactants do not prevent the bacteria from developing a slime layer, they can interfere with the "stickiness" of the slime and prevent the bacteria from sticking together. Additionally, excessive shearing caused by coarse air diffusers causes floc to break apart.

Operational Parameters That Affect Bacteria

Food-to-Microorganism Ratio

The food-to-microorganism (F/M) ratio compares the amount of available food with the number of microorganisms in the aeration basin. A high F/M ratio means there is a great deal of food and only a few microorganisms. When the F/M ratio is high, the bacteria are active and dispersed and multiply rapidly. Remember, in these conditions, they do not form floc. Operating the wastewater treatment system with a high F/M ratio will most likely yield a turbid effluent and a poor-settling sludge. A low F/M ratio means the opposite. This means there are many microorganisms with only a limited amount of food. Only when food is limited do bacteria begin to develop a thicker slime layer, lose their motility, and clump together to form floc. However, if the F/M ratio is too low, bacteria will starve of food and nutrients, and this can cause other operational problems we will discuss later.

How is the amount of available food determined? The F/M ratio should be determined by using the organic load entering the aeration basin in relation to the amount of microorganisms in the basin. The amount of food is measured using the BOD test. The BOD test is an indirect measure of the amount of food. To more accurately measure the amount of food available to microorganisms, BOD should be measured using the wastewater that directly enters the aeration basin. If primary treatment is available, the primary effluent should be used. As microorganisms consume organic material in wastewater, they use oxygen. The BOD test measures the amount of oxygen microorganisms require as they consume the available food. If a lot of food is available, the demand for oxygen will be high. Conversely, if little food is available, the demand for oxygen will be low. One drawback to this test is that the sample has to be incubated for five days. The concentration of oxygen in the water is measured before and after the five-day period. The difference is the amount of oxygen the microorganisms required as they consumed the food in the sample during that time. In other words, there is a direct correlation between the amount of oxygen used and the amount of organics consumed.

How is the concentration of microorganisms determined? Generally, the mixed liquor volatile suspended solids (MLVSS) concentration is used to determine the concentration of microorganisms in the aeration basin. Remember, the mixed liquor in the aeration basin is a mixture of microorganisms and wastewater. A sample of mixed liquor is poured and captured onto a filter that is dried to determine the dry weight. This dry weight is the concentration of mixed liquor suspended solids. The filter is then heated to about 550°C to burn off all the organic solids. The percentage that is burned off is called the volatile solids concentration, or MLVSS. This volatile portion is assumed to be mostly microorganisms. So, the amount of microorganisms is determined by measuring the MLVSS concentration.

How is the F/M ratio determined? The F/M ratio is the amount of food measured in pounds (lb) entering the aeration basin (as determined by the BOD test) divided by the pounds of microorganisms (as determined by the MLVSS test):

$$\frac{\text{lb BOD (entering the aeration basin)}}{\text{lb MLVSS (in the aeration basin)}} = \text{F/M ratio}$$

The F/M ratio affects the performance of the microorganisms in the activated sludge. Understanding how the F/M ratio affects the microorganisms will give the operator a valuable process control tool. There is not an ideal F/M ratio that will work for all activated sludge treatment systems. Generally, a lower F/M ratio (0.05–0.5) is acceptable.

Sludge Age

Sludge age refers to the average age of the sludge or microorganisms in the activated sludge system (aeration basin and clarifiers). For the purpose of this book, sludge age refers to the average time the microorganisms remain in the aeration basin, where bacteria come in contact with the incoming wastewater as a food source. There is an initial adjustment period during which the bacteria assess the environment and develop the enzymes that will be required to break down the nutrients they have detected in the wastewater. Once this occurs, they begin to consume the nutrients and multiply at a logarithmic rate. The bacteria are motile and active and do not clump together to form floc. Only by remaining in the basin long enough to deplete most of the nutrients do the bacteria begin to form the slime layer necessary to form floc. Enough bacteria must be allowed to remain in the aeration basin long enough to be exposed to conditions in which food is limited. If the sludge is too young, this will not occur. On the other hand, if the sludge is too old, floc will begin to break apart and yield a turbid effluent.

REFERENCE

Bruslind, L. 2017. "Microbial Growth." In *Microbiology*. Corvallis, Ore.: Oregon State University. https://bio.libretexts.org/Bookshelves/Microbiology/ Microbiology_(Bruslind)/09%3A_Microbial_Growth (accessed June 7, 2024).

Chapter 5

Protozoa

Bacteria are the primary agents in the removal of organic nutrients from wastewater, but they are boring to observe under the microscope (especially with the microscopes most operators can afford). Protozoa, on the other hand, are larger (5 μm to 1 mm) and come in a variety of shapes with a wide range of behaviors, and they are much more interesting to watch. Protozoa make up about 4% of the microorganisms in activated sludge. Although bacteria do most of the work of removing organic waste, protozoa also play a critical role in the treatment process by removing and digesting free-swimming dispersed bacteria and other suspended particles. Remember, not all bacteria are floc formers. Many of them remain suspended in the water after the floc formers begin amassing to form floc. Protozoa feed on the free, dispersed bacteria. This improves the clarity of the wastewater effluent.

WHY BOTHER LOOKING AT PROTOZOA?

The types of protozoa present give us some indication of treatment system performance. In a steady-state system, there is a natural progression in the dominance of the different protozoan species; activated sludge systems are far from steady state. The type and concentration of organic nutrients constantly change, as do factors such as temperature, amount of available oxygen (O_2), and food-to-microorganism ratio. Operators should not rely solely on the microscopic identification of protozoa. However, the relative dominance of the different protozoa species does give us an indication of treatment system conditions. Many times, a sudden change in the number and type of protozoa in the system can predict problems if adjustments aren't made. How protozoa move, feed, and compete for food will determine which species will dominate at any given time.

TYPES OF PROTOZOA

Most protozoa are aerobic, and mostly all of them can be found in areas with large water bodies. There is an enormous diversity in types of protozoa, and they are extremely varied in appearance and lifestyle. Some of the smaller protozoa, like bacteria, take in soluble nutrients by absorption through the cell membrane. Others have specialized structures or mouthlike openings and feed on other microorganisms such as bacteria and algae or other solid matter. Protozoa are also single-celled microorganisms and are usually classified based on how they move. For the purpose of studying activated sludge, we will classify them as follows:

- Amoebae
- Flagellates
- Ciliates
  - Free-swimming ciliates
  - Crawling (grazing) ciliates
  - Sessile (stalked) ciliates

The wastewater operator need not know how to precisely identify all the species of protozoa. It is helpful, however, to classify them into one of the five categories listed here because each species of protozoa dominates under different conditions.

Amoebae

Amoebae are the most primitive type of protozoa. They feed on small organic particulates but will also eat algae, bacteria, other small protozoa, or anything that will hold still enough. The amoeba is not just a shapeless blob waiting for food to draw near. It can sense a nearby concentration of food and will be drawn to it. Although the amoeba can absorb soluble nutrients, it is more drawn to particulate (solid) matter. Amoebae contribute very little to the overall treatment of wastewater. They move slowly and cannot compete with bacteria or other protozoa for food. Amoebae can only dominate early in the treatment process, when the amount of food is quite high. They begin to die as the amount of food decreases and the competition for food increases. Thus, when a sample is collected at the discharge end of the aeration basin (where most of the food has been removed), we should not expect to see large numbers of amoebae present in the sample.

Identifying Amoebae Under the Microscope

Amoebae are relatively clear and move slowly. Sometimes they do not move at all. They can sit dormant for hours. They do very little to bring attention to themselves, which makes observing them under the microscope difficult. Amoebae should be observed under high power with a phase contrast condenser with as little light as possible. Amoebae can be categorized into two types: the naked amoeba and the testate (shelled) amoeba. The naked amoeba is most commonly seen in science books and depicted in horror movies. It moves slowly by extending lobe-like projections called pseudopodia. Whenever it senses nearby food, it begins to extend its pseudopodia until they have completely enveloped the food and form a cavity or vacuole. Enzymes are then secreted into the vacuole to break down the food into soluble form so it can be absorbed into the cell. The vacuole will move slowly down the body of the cell and decrease in size as the nutrients are broken down and are absorbed into the cell.

Naked amoeba. There are several different species of naked amoebae. Naked amoebae are completely exposed and vulnerable to environmental conditions. Their internal structures and pseudopodia are clearly seen. *Amoeba proteus* is the species most likely to be seen when looking for amoebae. *Saccamoeba, Amoeba radiosa*, and *Korotnevella* are also naked amoebae commonly observed in activated sludge (Figure 5-1).

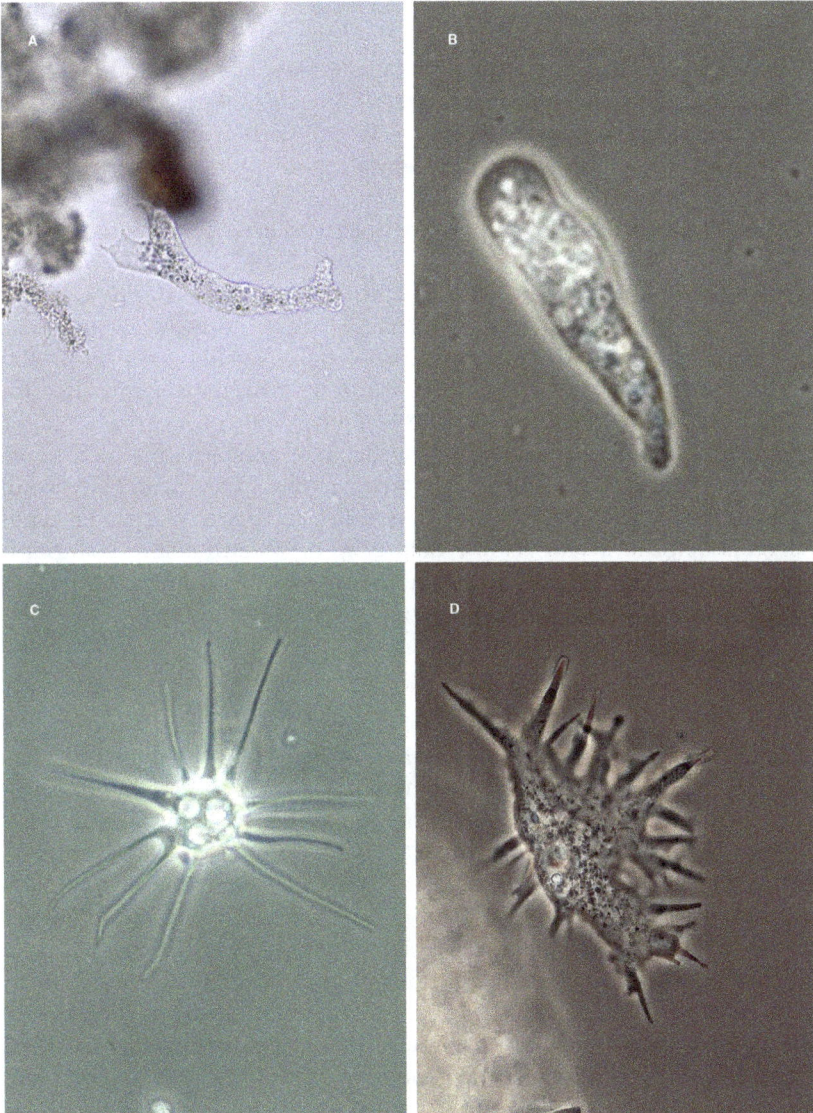

Figure 5-1   Naked amoebae: *A. proteus* (A), *Saccamoeba* (B), *A. radiosa* (C), and *Korotnevella* (D) (all 100× magnification, oil immersion; phase contrast)

Testate (shelled) amoebae. Testate amoebae have a test or "shell." They often are called "shelled amoebae." They can produce the shell either by secreting it or by building it from particles collected as they travel. Unlike naked amoebae, shelled amoebae are protected from environmental conditions and can survive in conditions that would otherwise be harmful. The way the test is constructed is a useful way of identifying the testate amoeba. Testate amoebae are sometimes difficult to identify. Many are mistaken for a piece of organic material. Careful and patient observation is required to properly identify them.

There are several different types of shelled amoebae commonly observed in activated sludge. Two species of *Arcella* are commonly observed: *Arcella discoides* and *Arcella vulgaris* (Figure 5-2). The shell provides protection and enables them to survive in environments that are too harsh for the naked amoebae.

Other shelled amoebae commonly observed are *Centropyxis*, *Euglypha*, *Difflugia*, and *Assulina* (Figure 5-3).

Heliozoa and *Plagiophrys* are also amoebae (Figure 5-4). They look quite different from the others. Heliozoa have a round body with many ridged spikes (axopodia) radiating outward from the cell body. The axopodia are sticky and have a surface that is in constant motion. Food items stick to the axopodia and are transported to the cell body by the flow of cytoplasm. *Plagiophrys* also have a round body, but the axopodia are thin and branched.

Operators need not identify each type of amoeba by name. Determining the number of amoebae in the sample compared with the other types of protozoa is helpful. It will also be helpful to be able to distinguish between naked and shelled amoebae and to note any significant increase in shelled amoebae in comparison. We will discuss this later in the book.

Flagellates

Flagellates are so named because they possess whiplike structures that help pull them through the water. A flagellate is usually quite small and has one or two flagella. It has a tough outer membrane and can ingest food through a specialized mouth called a cytostome. Bacteria also have flagella, but they are much different from that of the flagellate. A bacterial flagellum is like a small device that moves by rotating the shaft (the part where the flagellum is attached to the cell). In a bacterium, the flagellum works like a little propeller. In flagellates, the flagella are composed of small tubules of protein called microtubules. Their flagella not only propel them through the water but are also used for feeding. Amoebae prefer to feed on small particulates; flagellates thrive in waters that are rich in organic nutrients, especially where there is much decay. Some groups of flagellates feed on bacteria and small algae. However, many that thrive in the activated sludge treatment system seem to feed primarily on soluble organic nutrients, so they can only dominate early in the treatment process, when the amount of nutrients is quite high. Flagellates compete with bacteria for soluble nutrients; therefore,

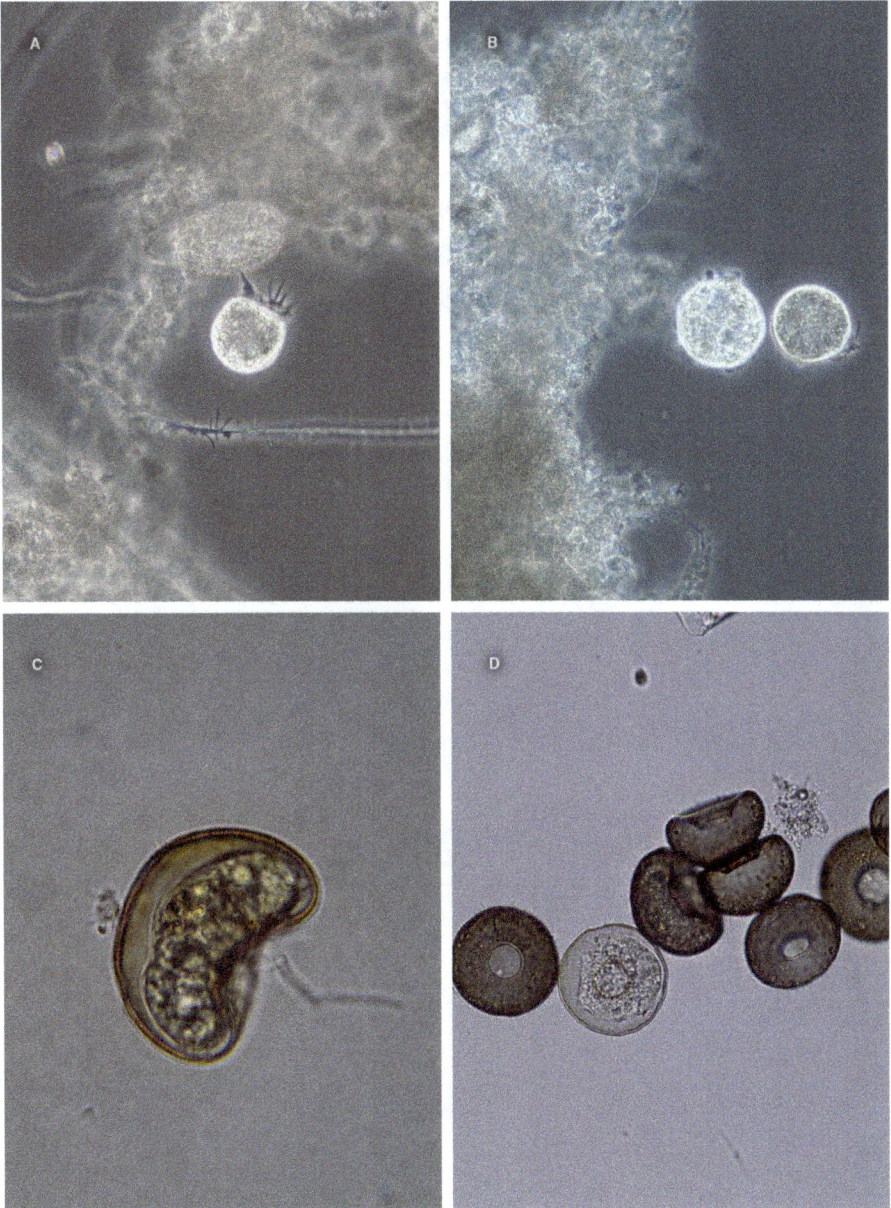

Figure 5-2   Shelled amoebae: *Arcella vulgaris* (A and B) and *Arcella discoides* (C and D) (all 40× magnification; phase contrast)

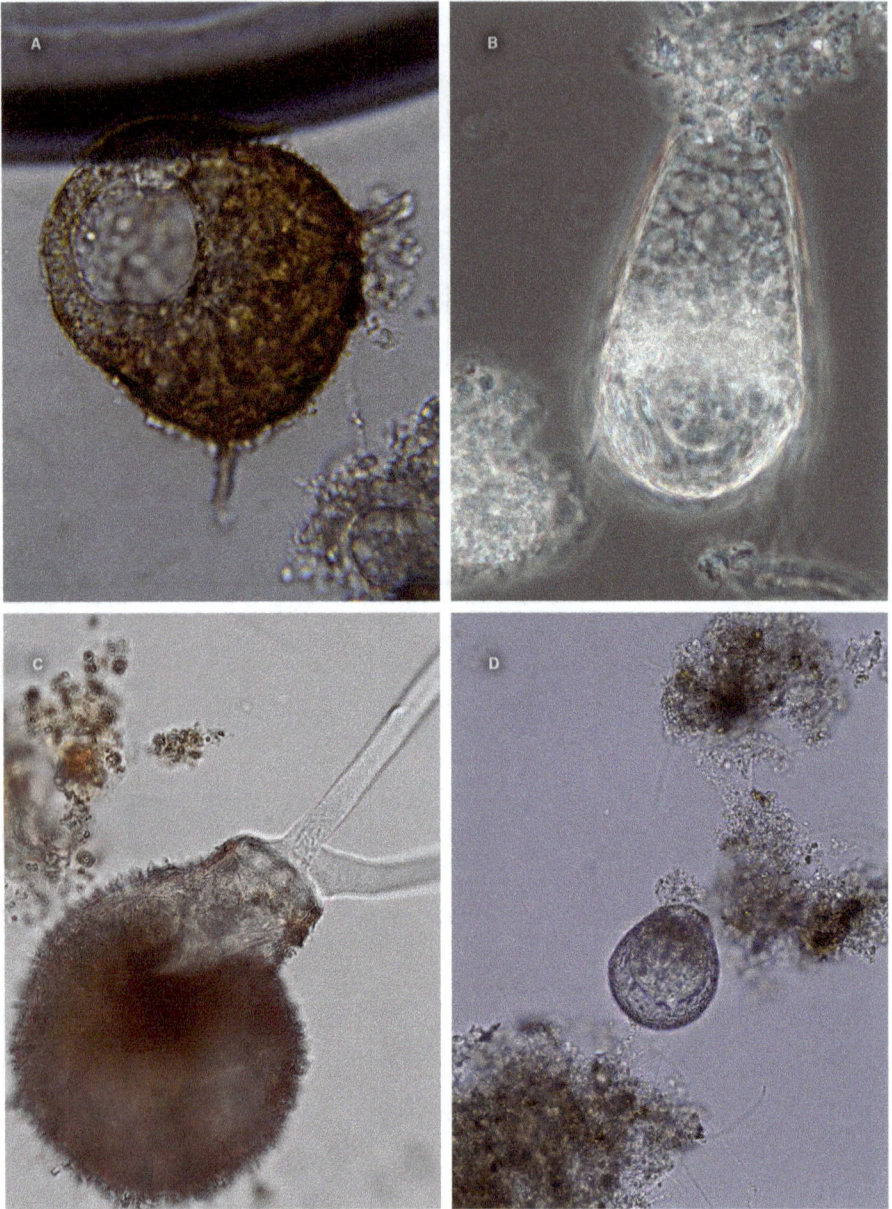

Source: Image C printed with permission from Glymph-Martin 2024b

Figure 5-3   Shelled amoebae: *Centropyxis* **(A),** *Euglypha* **(B),** *Difflugia* **(C), and** *Assulina* **(D)
(all 100× magnification; oil immersion; phase contrast)**

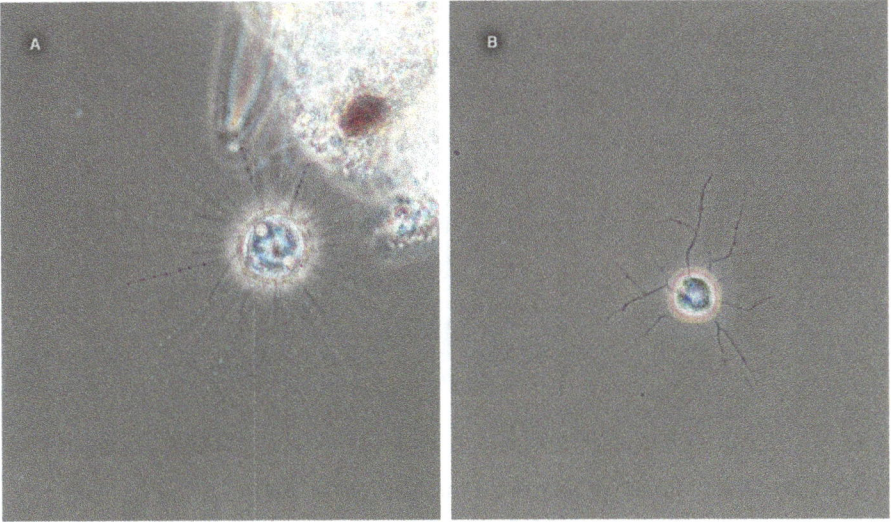

Figure 5-4   Heliozoa (A) and *Plagiophrys* (B) (both 100× magnification; oil immersion; phase contrast)

they can only dominate when the bacteria population is still relatively small. While bacteria are busy acclimating to the environment, flagellates are free from any real competition for food. However, after the bacteria's initial lag phase comes an accelerated growth phase. Bacteria begin to multiply and rapidly consume soluble food. The multiplication rate for bacteria is 15 times faster than that of the flagellates, so they are not without competition for long. Flagellates are also "saprophytic," which means they can derive nutrition from decaying organic matter. So, often after a toxic event, flagellates can feed off other dead and decaying microorganisms. An abundance of flagellates in the treatment system indicates there is a significant amount of soluble organic compounds still remaining in the wastewater.

Identifying Flagellates Under the Microscope

Flagellates can be seen in a variety of shapes and sizes (Figure 5-5). They are the most widespread of the protozoa. Most are relatively small and easily distinguished compared with the other protozoa commonly found in activated sludge. *Bodo* is the most common flagellate found in activated sludge. Its short, capsule-shaped body is often seen spiraling through the fluid. The whiplike structure, called a flagellum, is used for locomotion and feeding. Other flagellates such as *Peranema* are much larger. Their bodies can undergo flexible, shape-changing movements.

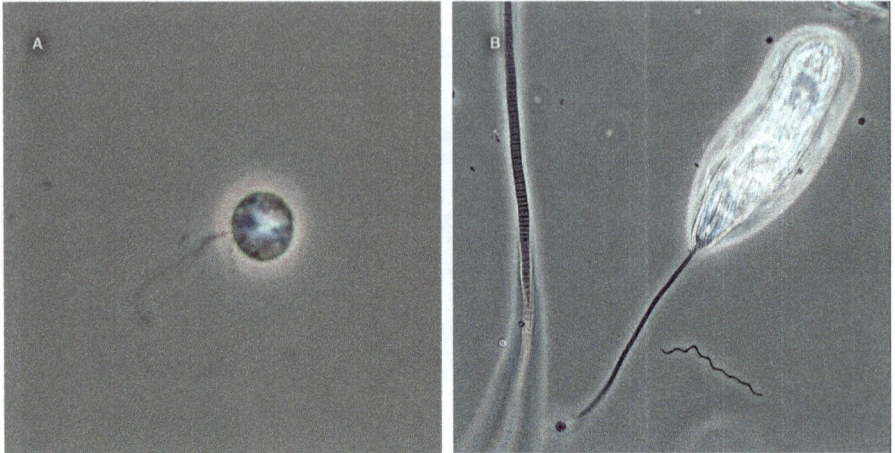

Source: Images A and B printed with permission from Glymph-Martin 2024a

Figure 5-5   Flagellates: *Bodo* **(100× magnification; oil immersion; phase contrast) (A) and** *Paranema* **(40× magnification; phase contrast) (B)**

Ciliates

Ciliates are either completely or partially covered with short, dense, hairlike structures called cilia, which get their meaning from the Latin word for eyelash. The cilia are used primarily for locomotion. Although ciliates actively capture their prey, the cilia are arranged in such a way that they can be used to generate water currents that will sweep the food into the ciliate's mouth. Although they are single-celled microorganisms, they still perform complex biological functions. Some ciliates can reach 2 cm in length.

Ciliates feed mostly on bacteria, algae, and yeast, so they do not contribute to the overall treatment of the wastewater by actively removing organic material. They do contribute to the overall clarity of the water by removing suspended bacteria, algae, and other small microorganisms. Bacteria multiply as they consume organic material; ciliates multiply as they consume bacteria. An abundance of ciliates usually indicates that most of the organic material has been removed from wastewater. About 7,500 species of ciliates are generally classified based on the way the cilia are arranged. For the purpose of studying activated sludge, we will look at three categories of ciliates. They are categorized based on their feeding habits and their ability to compete for food. We will discuss free-swimming, crawling, and sessile ciliates.

Free-Swimming Ciliates

These are generally uniformly covered with cilia. They move by swimming and, in many instances, spiraling through the water. Some ciliates, such as the *Paramecium*, can back up and change directions if they encounter a solid object. Because free dispersed bacteria are their principal food, free-swimming ciliates begin to dominate when the numbers of bacteria are quite high. Bacteria reach their peak after most of the soluble nutrients have been depleted. Flagellates cannot compete with bacteria for organic nutrients, but free-swimming ciliates do not have to compete with the bacteria. Free dispersed bacteria give them the advantage. In the presence of an abundant food source, bacteria are motile and multiply actively. Free-swimming ciliates move quickly through this environment and consume as many bacteria as they can. Once the organic nutrients become limited, the bacteria will no longer swim freely but will begin to lose the flagella, form a sticky slime layer, and clump together to form floc. It is difficult for free-swimming ciliates to feed on bacteria amassed in the floc. So, once the bacteria begin to form floc, the population of free-swimming ciliates will begin to die off.

Identifying free-swimming ciliates under the microscope. Free-swimming ciliates use their cilia primarily for locomotion. They move quickly through the fluid and sweep bacteria into an oral groove (mouth). They are easily distinguished from the crawling ciliates and sessile ciliates because they move quickly and generally hang out in the fluid. Something that swims quickly past your field of view in the microscope is probably a free-swimming ciliate. There are hundreds of species of free-swimming ciliates. Some that are commonly observed in activated sludge are *Litonotus, Acineria, Coleps, Blepharisma* (Figure 5-6), *Glaucoma, Spirostomum, Tetrahymena,* and *Trachelophyllum* (Figure 5-7).

Litonotus is known as a graceful swimmer. It is rarely observed within the floc. *Acineria* are uniquely adaptable to fluctuating oxygen levels in wastewater. *Coleps* are known as the armored tanks of the protozoa world because their bodies are protected by many small plates and spines. They are also scavengers, feeding on the dead remains of other protozoa as well as bacteria. *Blepharisma* is known for its pinkish color, and *Glaucoma* frankly reminds me of a robot vacuum cleaner.

Spirostomum are some of the largest species of protozoa. They hold the record for the fastest body contractions of any living cell, contracting their length to 25% of their normal size in 6–8 ms. Interestingly enough, *Tetrahymena* can produce, store, and secrete melatonin. *Trachelophyllum* are predatory. Not only do they feed on bacteria, but they also feed on smaller protozoa.

Knowing the names of individual free-swimming ciliates is not important. It is, however, important that the operator is able to identify that the microorganism is a free-swimming ciliate. If it tends to hang out and swim around in the bulk fluid or even swims pretty fast, it is probably a free-swimming ciliate, and that is all the operator needs to know. An abundance of free-swimming ciliates is an indication that there are a lot of bacteria present (because that is their primary food). A lot of bacteria means most of the organic compounds have been removed.

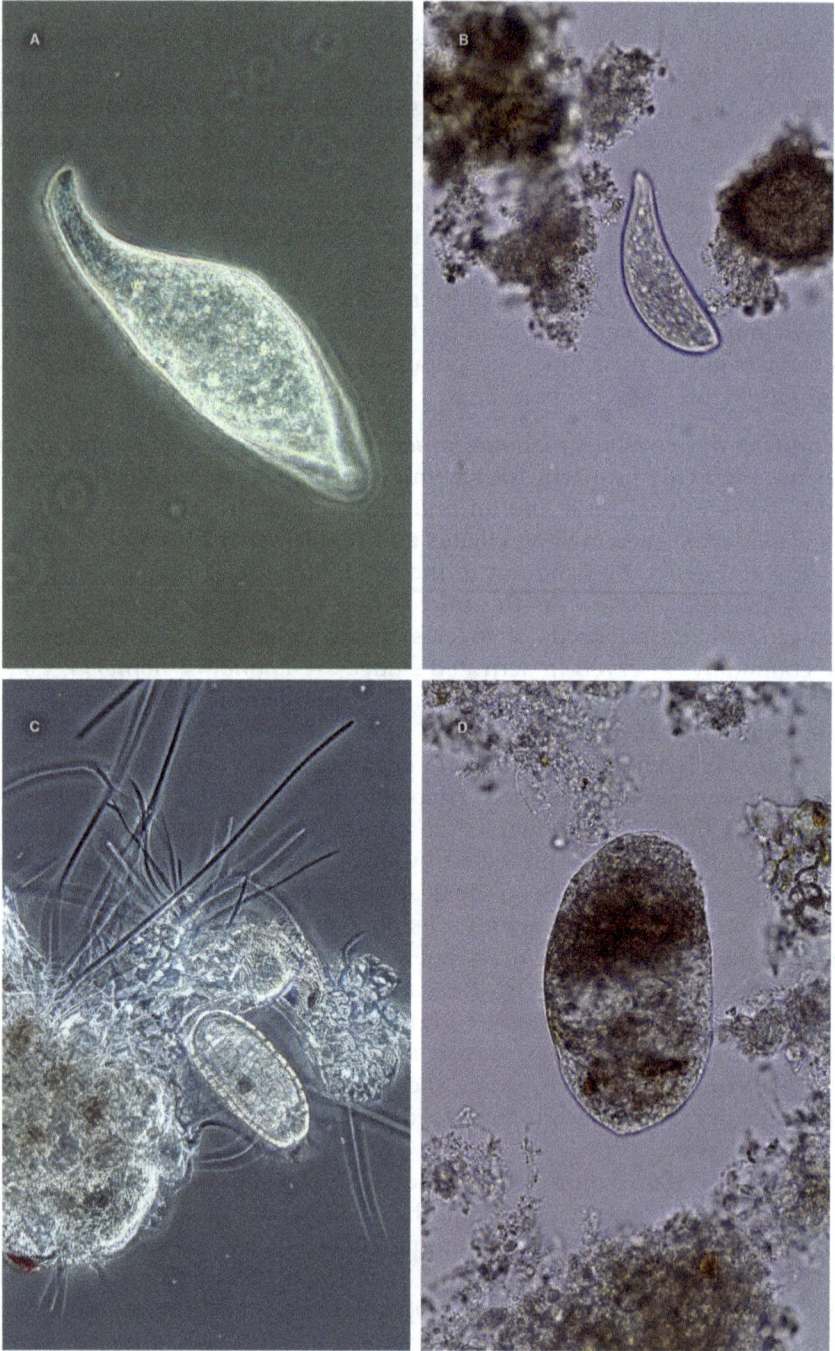

Figure 5-6 Free-swimming ciliates: *Litonotus* **(A),** *Acineria* **(B),** *Coleps* **(C), and** *Neobursaridium* **(D) (all 40× magnification; phase contrast)**

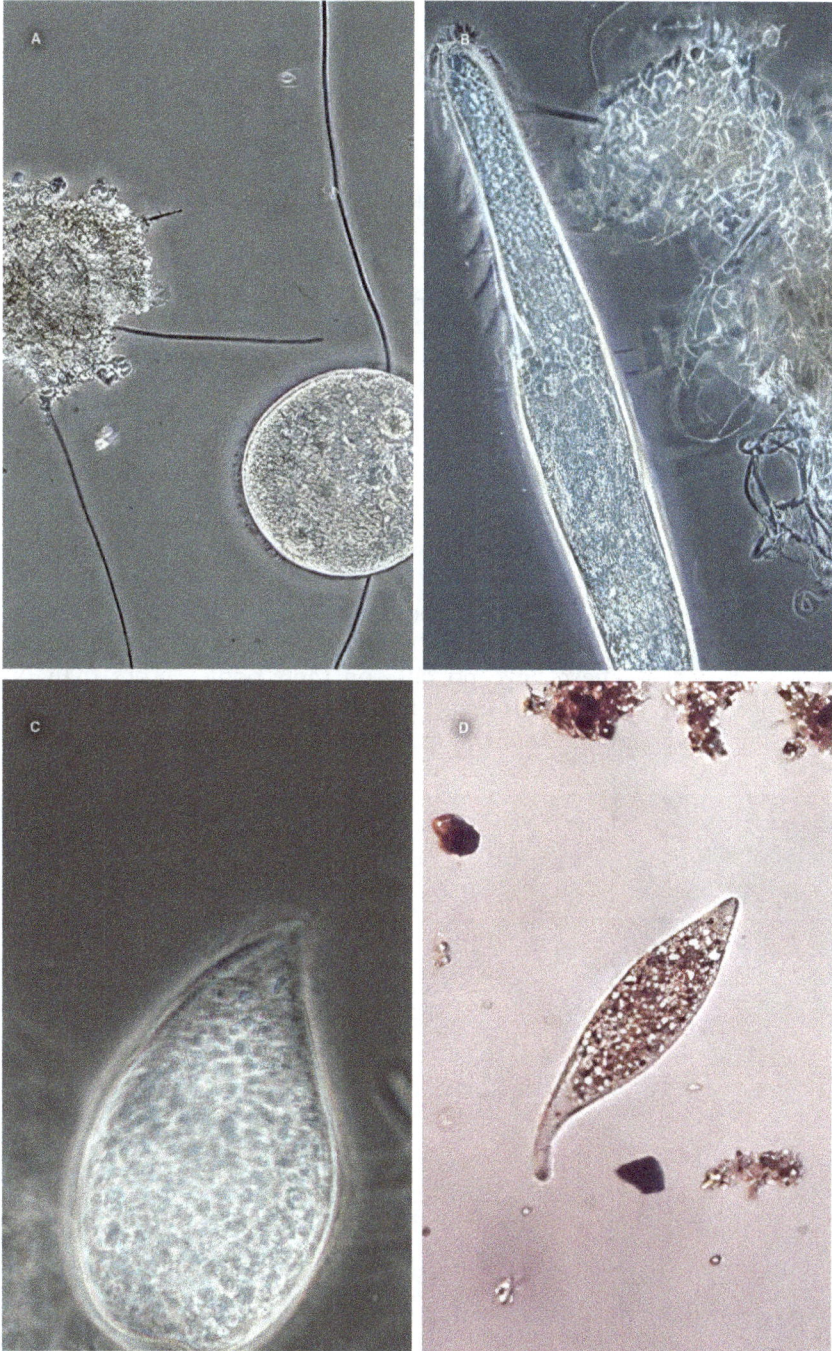

Figure 5-7 Free-swimming ciliates: *Glaucoma* (A), *Spirostomum* (B), *Tetrahymena* (C), and *Trachelophyllum* (D) (all 40× magnification; phase contrast)

Crawling (Grazing) Ciliates

These are common in activated sludge. Their dominance in the system indicates good treatment conditions. Crawling ciliates begin to gain dominance after most soluble nutrients have been removed. During this time, most of the dispersed bacteria have begun to amass to form floc.

Identifying crawling ciliates under the microscope. Crawlers have cilia that are organized into structures called cirri, which are several cilia joined together to form a tuft or leg. Instead of swimming, the crawling ciliate has cirri on its underside that function as coordinated legs that allow them to walk on surfaces. They use these legs to browse or crawl along solid floc particles and pick at any straggling bacteria they can find. Crawlers are poor swimmers. If left in the fluid, they grope around, turn in circles, and try to touch something solid so they can climb aboard and begin to crawl (Figure 5-8).

Sessile Ciliates

Sessile means "stationary." Some sessile ciliates are attached to stalks, thus the term "stalked ciliates." The stalk is made up of a proteinaceous material that the ciliate "head" (termed *zooid*) secretes. At the base of the stalk, there is an adhesive disc or secretion that secures the stalk to solid surfaces like floc particles. These have no cilia on their bodies other than those that fringe around the mouth ends. The cilia are used to create a current that brings food particles into the mouth. They feed mostly on suspended

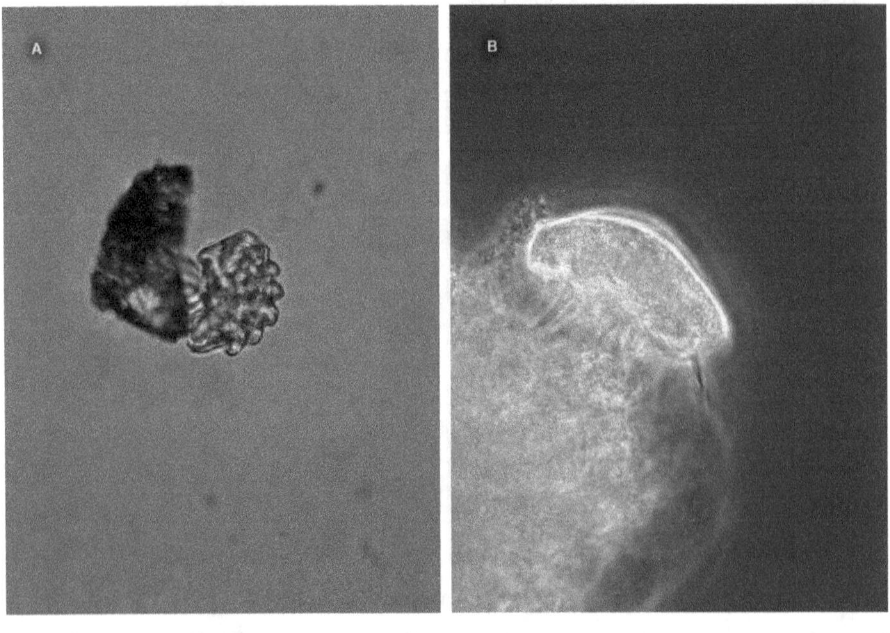

Figure 5-8 Crawling ciliates: *Aspidisca* (100× magnification; oil immersion; phase contrast) (both A and B)

bacteria, algae, smaller protozoa, or anything that will fit into their mouth. They rarely swim freely but can be found attached to almost anything. Some stalked ciliates grow as separate stalks; others branch into colonies that can be made up of 100 or more organisms. By the time stalked ciliates begin to dominate in the activated sludge, the bacteria have removed most of the organic nutrients; most of the free dispersed bacteria have been removed by the free-swimming ciliates or have developed into floc. The stalked ciliate's advantage is that it doesn't have to go looking for food. Instead, it can anchor itself while it uses its cilia to create a current that brings the food to its mouth.

Identifying stalked ciliates under the microscope. The stalked ciliate *Vorticella* grows on a single stalk and does not grow in colonies. They are sometimes found in large groups with separate stalks. The stalks are tubular structures with central muscle fibers that cause the ciliate to contract whenever it is disturbed (Figure 5-9). Once the disturbance has passed, the ciliate will slowly extend itself again. Stalked ciliates can and will detach themselves from their stalks whenever conditions become unfavorable and swim freely to look for a more favorable environment where they can generate new stalks. Whereas *Vorticella* has a single stalk with a single head, some stalked ciliated grow in colonies with several heads attached to branching stalks.

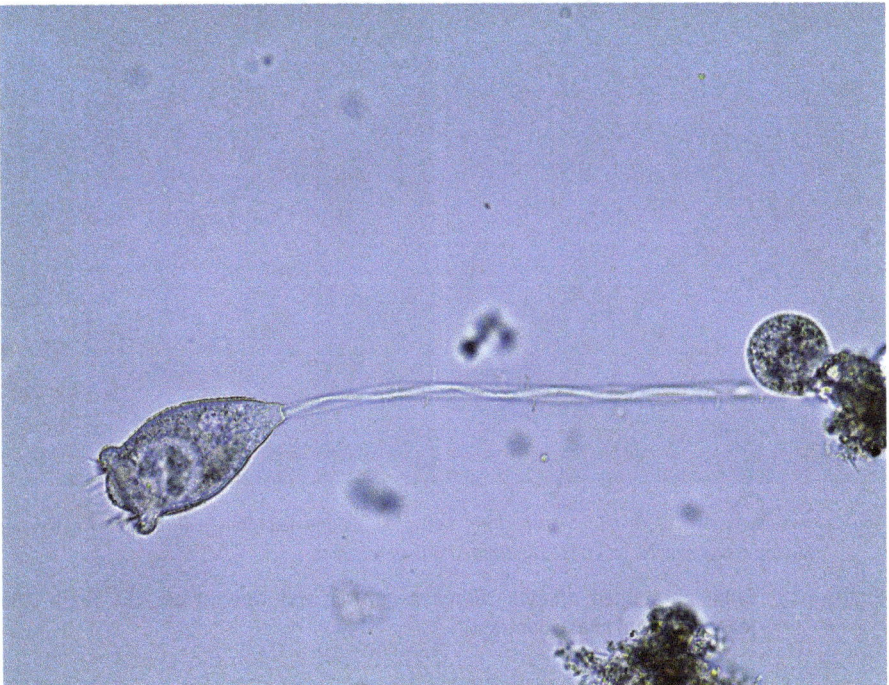

Figure 5-9 Stalked ciliate: *Vorticella* (40× magnification; phase contrast)

Carchesium, which looks a lot like *Vorticella*, is a colonial stalked ciliate. Several zooids, or heads, are attached to branching stalks. Each stalk has its own contractile muscle that contracts independently of the others and of the main trunk, which also has a contractile muscle. A *Carchesium* colony can contain hundreds of heads. The combined action of all the heads can consume a large quantity of suspended bacteria and algae. *Epistylis* is another type of colonial stalked ciliate. Similar to *Vorticella* and *Carchesium*, it has an inverted bell-shaped head, but it is mounted on a thick branched non-contractile stalk. *Opercularia* also has a branched noncontractile stalk but grows in much smaller colonies that consist of 3–50 individual heads (Figure 5-10 and Figure 5-11).

Although rarely seen with cilia, suctorians are also stalked ciliates. In their mature sessile form, suctorians do not have cilia for locomotion or feeding. However, during their early developmental stages, they possess cilia and are motile. Once they mature, they lose their cilia and become sessile. Much like its name, the suctorian has lots of long tubular tentacles with sucker-like tips that suck the body fluids of its prey (Figure 5-12 and Figure 5-13).

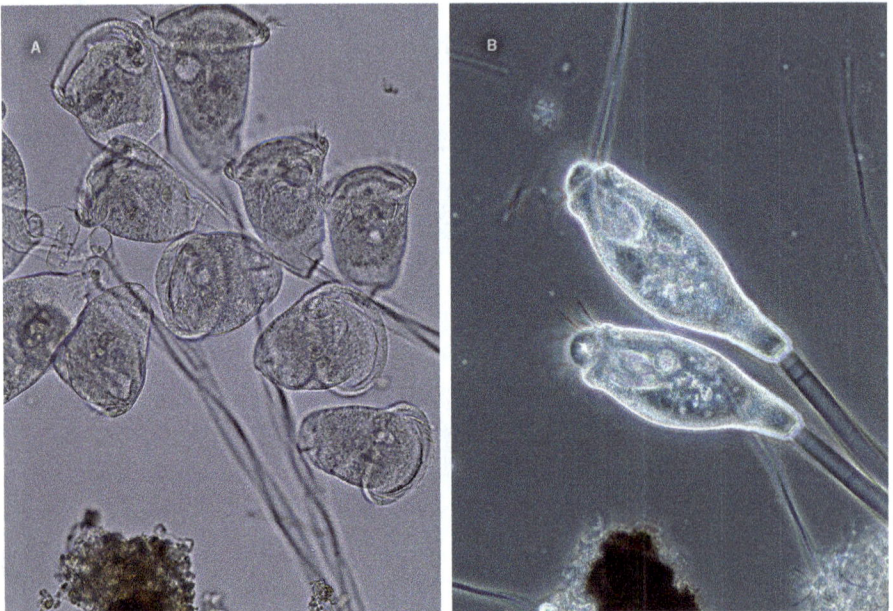

Source: Images A and B: Glymph-Martin, Toni, Activated Sludge Microbes - Protozoa. Wastewater Microbiology Solutions LLC (2024). Printed with permission.

Figure 5-10 Colonial stalked ciliates: *Carchesium* (A) and *Opercularia* (B) (both 40× magnification; phase contrast)

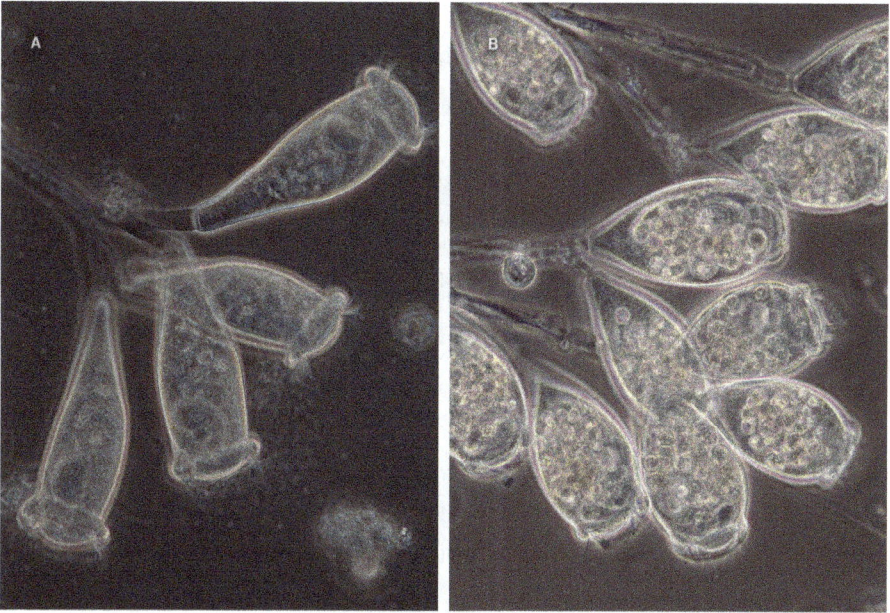

Figure 5-11   Colonial stalked ciliates: *Epistylis* (A) and *Zoothamnium* (B) (both 40× magnification; phase contrast)

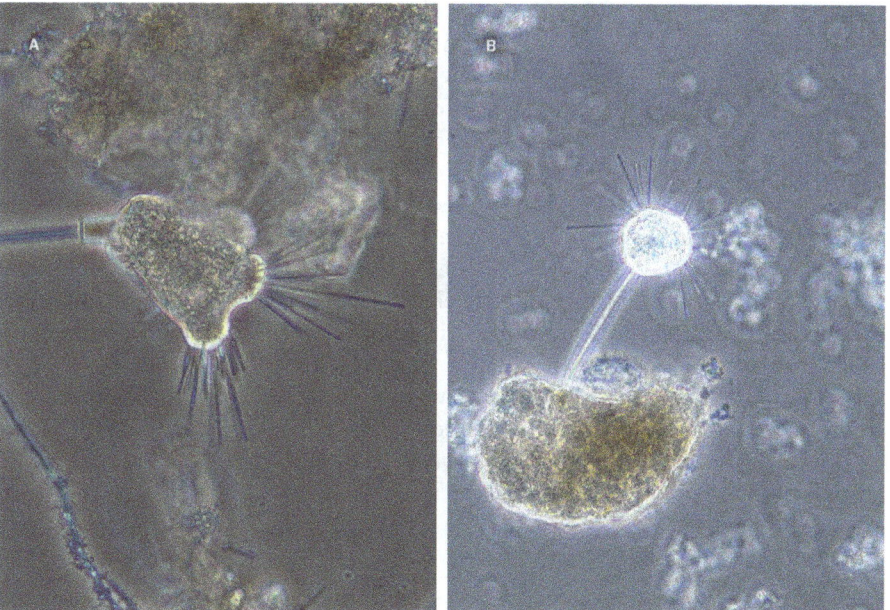

Figure 5-12   Suctoria: *Acineta* (40× magnification; phase contrast) (A) and *Podophrya* (100× magnification; oil immersion; phase contrast) (B)

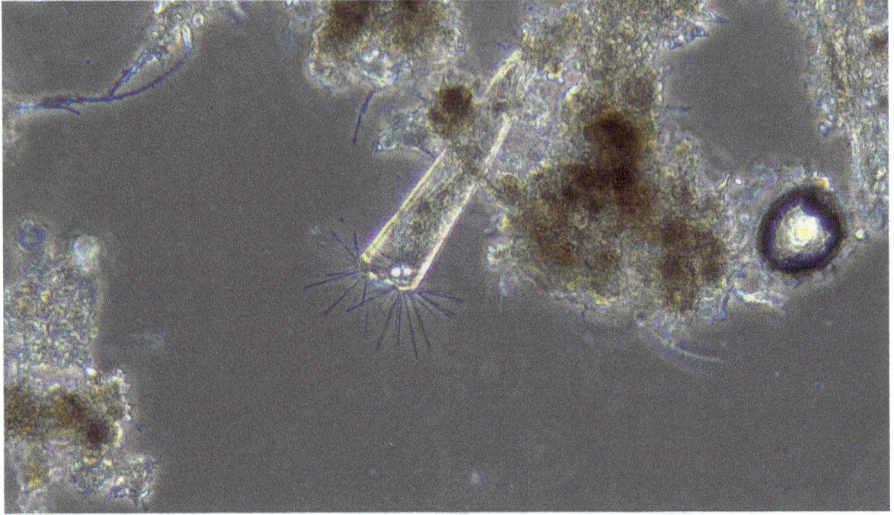

Figure 5-13   Suctoria: *Dendrosoma* (40× magnification; phase contrast)

Not all sessile ciliates have a stalk. The *Stentor*, for example, is one of the largest protozoa in water. It is a trumpet-shaped organism that is covered with short cilia. A crown of longer cilia, used to aid in feeding, surrounds the rim of the trumpet (Figure 5-14). Stentors can either swim freely or form colonies by secreting a sticky gelatinous substance that anchors them together. Any disturbance will cause the organism to contract into what looks like a blob.

Some sessile ciliates are "lorica-forming" protozoa. I like to call them "tube-dwellers." These ciliates form a protective outer shell that can be composed of silica, calcium carbonate, or organic substances. They serve as a protective covering against other predators and even from an unfavorable environment (Figure 5-15). *Vaginicola* is a tube-dwelling ciliate enclosed in a vaselike sheath. Generally, a pair of these organisms lives together in one tube. *Thuricola* is another tube-dwelling ciliate and is typically observed as one organism to a tube. Both of these microorganisms can retract completely into their tubes for protection when disturbed.

Although protozoa play a secondary but critical role in wastewater treatment, the relative abundance of these types of microorganisms can be an indicator of the conditions of the activated sludge. Understanding their feeding mechanisms and their competitors will help us understand their dominance in the system.

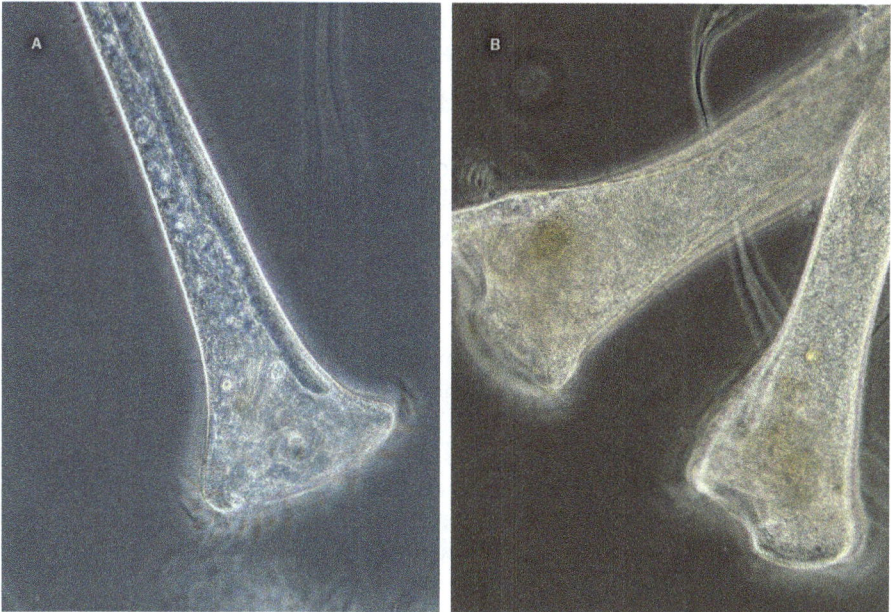

Source: Image A: Glymph-Martin, Toni, Activated Sludge Microbes - Protozoa. Wastewater Microbiology Solutions LLC (2024). Printed with permission.

Figure 5-14 *Stentor* (40× magnification; phase contrast) (both A and B)

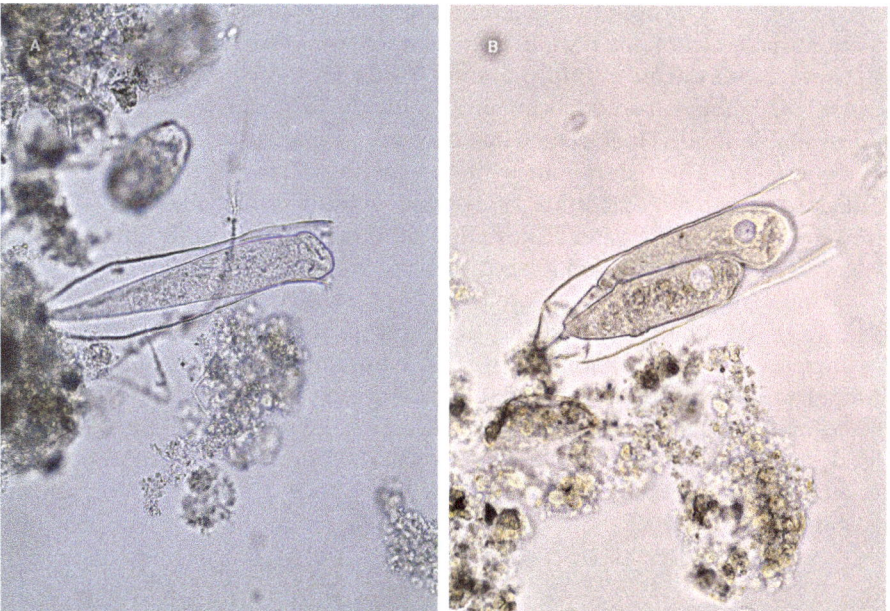

Figure 5-15 Lorica-forming (tube-dwelling) sessile ciliates: *Thuricola* (A) and *Vaginicola* (B) (both 40× magnification; phase contrast)

WHAT IF AMOEBAE ARE DOMINANT IN A SAMPLE OF ACTIVATED SLUDGE?

If a sample is collected from the discharge end of the aeration basin and if enough time was allowed in the basin, most of the organic material should have been removed with little remaining. Amoebae move slowly and have difficulty competing for food when the levels are very low. Some amoebae will always be present in the activated sludge. However, if large numbers of amoebae are present in a sample collected at the end of the aeration basin, this suggests that there is still a significant amount of available food in the wastewater. This may indicate one or more of the following:

- A shock load of BOD has entered the treatment system: Sudden increase in food allows amoebae to compete and multiply.
- A large amount of particulates: The amoebae favor particulate food; bacteria prefer soluble nutrients. So, expect to see more amoebae after heavy rain, when particulates are washed in with the stormwater.
- Low dissolved oxygen (DO): Because amoebae are the most primitive of the protozoa and move slowly, they require less oxygen and can compete with other protozoa when oxygen is limited.

WHAT IF FLAGELLATES ARE DOMINANT IN A SAMPLE OF ACTIVATED SLUDGE?

Flagellates eat bacteria but prefer soluble organic material. They are also what is called "saprophytic," meaning they feed on dead or decaying organic matter. In the treatment system, they compete with bacteria for soluble food. Because bacteria can multiply every 20 min and flagellates multiply once every 5–20 hours, bacteria can grow quickly to compete with flagellates for available food. Thus, flagellates can only dominate when the food level is quite high. Their dominance in the activated sludge treatment system indicates significant amounts of soluble organic substances may still be present in the wastewater. They also dominate immediately after a toxic event because they feed on dead material.

Amoebae and flagellates are always present in activated sludge; however, they dominate early in the treatment process. As the bacteria begin to increase in number, soluble organic compounds are consumed rapidly. Amoebae have no built-in mechanisms that would allow them to compete for food. Although the flagellates are better equipped to move and search for food, they cannot compete with the logarithmic growth rate of the bacteria. In a continuously fed batch process, the dominance of these microorganisms is short-lived.

WHAT IF FREE-SWIMMING CILIATES ARE DOMINANT IN A SAMPLE OF ACTIVATED SLUDGE?

The dominance of free-swimming ciliates generally indicates an abundance of active bacteria. Although the presence of lots of bacteria suggests that much of the organic material has been removed (because they multiplied exponentially), this means it is still somewhat early in the treatment process, and the organic material has not been depleted to a level that would cause the bacteria to begin to form floc. If a sample collected at the end of the activated sludge process contains a dominance of free-swimming ciliates, the effluent will most likely be turbid and full of nonflocculated bacteria. Usually, very little floc is formed, and the solids will not settle well. Treatment is not considered complete until enough of the organic material has been consumed to cause the bacteria to form floc that will settle well in the final clarifier. Increasing the sludge age will give the bacteria enough time in the aeration basin to remove most of the organic material and to begin to form floc.

WHAT IF CRAWLING CILIATES ARE DOMINANT IN A SAMPLE OF ACTIVATED SLUDGE?

Once most of the organic material has been removed, the floc-forming bacteria are no longer freely dispersed but are clumped together to form floc. Free-swimming ciliates begin to lose their advantage. Crawling ciliates, however, begin to dominate because they can find food among the floc particles. An abundance of crawling ciliates means an abundance of floc, which means most of the organic material has been removed from the wastewater. It also indicates that the sludge age of the treatment system is adequate.

WHAT IF STALKED CILIATES ARE DOMINANT IN A SAMPLE OF ACTIVATED SLUDGE?

An abundance of stalked ciliates indicates that most of the organic material has been removed. As the sludge begins to age, the dominance of stalked ciliates changes from single stalks to colonial species. So, keep in mind, "the greater the number of heads, the older the sludge."

What About Stentors and Suctorians?

Stentors and suctorians rarely (and should never) dominate in a sample of activated sludge. As the sludge ages, these have a competitive advantage. Stentors are large and, like stalked ciliates, can create a current and draw the wastewater into their mouth. Because the stentor is larger and can fill up its whole body, it has a better chance to get the limited amount of available food. Suctorians, on the other hand, are not affected by the lack of bacteria or other organic material. They can feed on other protozoa by sucking them into their tentacles. So, increased numbers of these suggest an older sludge age.

FACTORS THAT AFFECT PROTOZOA

Numerous environmental factors influence the reproduction and death of protozoa. Factors such as nutrients, DO, and competition influence the abundance and types of microorganisms in the treatment system. Like bacteria, protozoa need oxygen. A lack of DO will severely limit the type and number of these microorganisms in the system. They can survive and reproduce in most temperatures that are common to activated sludge treatment systems. However, like most organisms, their metabolism is slower in colder temperatures. In fact, their metabolic activity doubles with every 10°C rise in temperature. Some can increase their growth rates by more than three times between 10 and 20°C. They grow best in ambient temperatures around 15–25°C. However, sudden changes in temperature can cause death, but with slow and gradual acclimation, protozoa can be made to survive temperatures outside their normal temperature ranges. Most municipal wastewater contains enough nutrients to support most protozoa.

PROTOZOA IN ACTIVATED SLUDGE

Typically, protozoa depend on dead and living bacteria for food, and bacteria in turn depend on the organic material in the wastewater for food. While few protozoa compete with bacteria for organics, most protozoa compete with one another for bacteria. Bacteria multiply much faster than protozoa. So, as long as organic material is available, bacteria will outcompete protozoa for the organics. Ciliated protozoa, on the other hand, compete with each other for bacteria. Their dominance in the treatment system depends on who can consume the most bacteria. As long as the bacteria are active and freely dispersed, free-swimming ciliates will have a competitive edge. They move quickly throughout the fluid, consuming bacteria at an enormous rate. But as the nutrient levels decrease and bacteria begin to form floc, crawling ciliates gain the competitive edge. They can graze along the floc and consume straggling bacteria. As food levels and the number of bacteria continue to decrease even more, stalked ciliates gain the advantage. Other ciliates must travel to find food, but stalked ciliates can create a current that will bring the food to them. As the competition for food increases, larger protozoa like stentors and predators such as suctorians gain the advantage.

REFERENCES

Glymph-Martin, T. 2024a. *Activated Sludge Microbes Poster 1*. Wastewater Microbiology Solutions.

Glymph-Martin, T. 2024b. *Activated Sludge Microbes Poster 2*. Wastewater Microbiology Solutions.

Chapter 6

Metazoans

Metazoans are multicellular organisms and include all animals except the protozoa. They are much more complex and range from bumblebees and spiders to fish and elephants. For the purpose of studying activated sludge, we will discuss some of the smaller microscopic metazoa and their role in the wastewater treatment system. Metazoans are larger than most protozoa and should be observed with the phase contrast condenser and no greater than the 40× objective. For some of the larger metazoa, it may be necessary to use the 10× objective.

Metazoa have very little to do with the removal of organic material from the wastewater. Although they eat bacteria, they also feed on algae and protozoa. A dominance of metazoa is usually found in longer-age systems—namely, lagoon treatment systems. Although their contribution to the removal of organic material in the activated sludge treatment system is small, their presence indicates treatment system conditions. Rotifers, gastrotrichs, nematodes, bristle worms, and tardigrades (water bears) are the most common metazoans found in the activated sludge treatment system.

ROTIFERS

The principal contribution of rotifers is to clarify the effluent by removing leftover bacteria, algae, or other smaller protozoa. Sometimes, as many as four types of rotifers may be present in the treatment system at one time. Rotifers should never dominate in the treatment system. They reach their peak after most of the nutrients have been removed from the wastewater. The presence of dead rotifers in a fresh sample of mixed liquor is a good indicator of wastewater toxicity. They are usually the first to be affected by toxicity in the wastewater.

Rotifers Under the Microscope

Rotifers come in a variety of shapes and sizes and are amazingly complex. The word rotifer comes from the Greek word meaning "wheel-bearing animal." The wheels of the rotifer are actually a structure called the corona, which is covered with cilia. The corona moves constantly and brings food into the rotifer or provides a means of locomotion. Some species appear to walk, whereas others secrete a sticky substance that enables them to attach to surfaces. A rotifer also has a unique structure called a mastax that has many muscles that control a set of jaws. The jaw structure grinds the food that is brought down by the action of cilia. The female rotifer is much larger than the male, whose life span is very short and whose only purpose for living is to fertilize the female. Figure 6-1 shows two different species of rotifers—*Rotaria* and *Epiphanes*. The species *Rotaria neptunia* is the most common and can stretch longer than a millimeter. *Epiphanes* is a very large rotifer. It can be found in almost all aquatic habitats.

Similar to amoebae, some rotifers like *Lecane* (Figure 6-2) have a shell. Most rotifers are 200–500 µm long.

GASTROTRICH

Gastrotrichs are also known as "hairy bellies" because they have cilia covering their underside or "belly." They are found in the same environments as rotifers and nematodes. Gastrotrichs feed on small particles, dead or alive, such as plant debris, bacteria, algae, or protozoa.

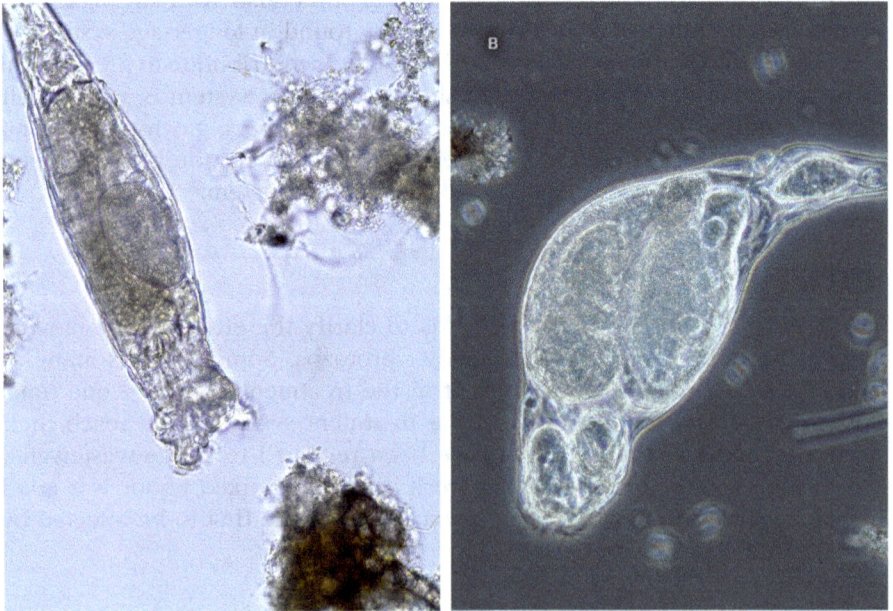

Figure 6-1 Rotifers: *Rotaria neptunia* (A) and *Epiphanes* (B) (40× magnification; phase contrast)

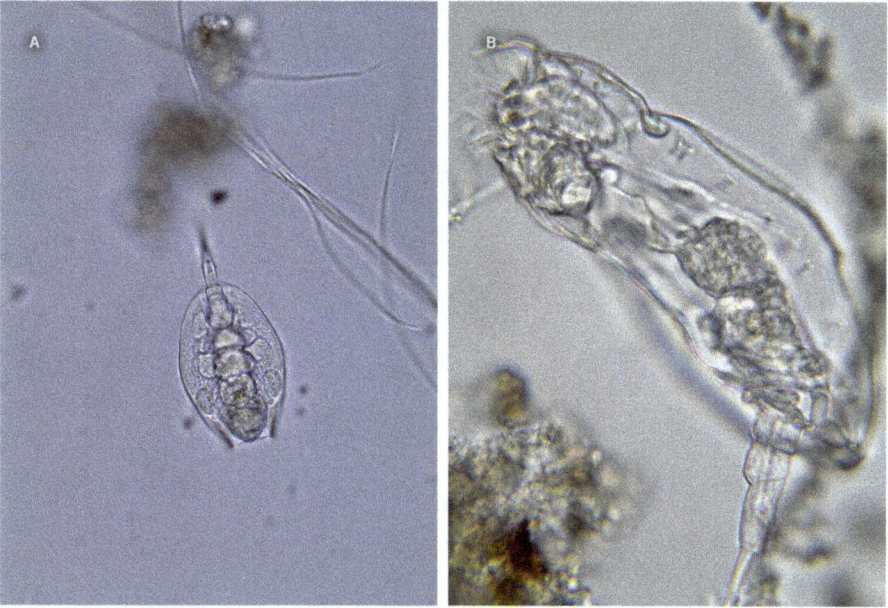

Figure 6-2 Shelled rotifer: *Lecane* (40× magnification phase contrast) (A); *Colurella* (100× magnification; oil immersion; phase contrast) (B)

Gastrotrichs Under the Microscope

Gastrotrichs appear somewhat flattened, with a swollen head and forked end (Figure 6-3). They are often mistaken for rotifers because of the forked end. However, the forked tail of the rotifer can open and close and is used for attachment, whereas the forked end of the gastrotrich is fixed and does not open and close.

NEMATODES

Nematodes have a simple structure, even though they are also multicellular. They possess a digestive, reproductive, and nervous system. Nematodes feed on bacteria, fungi, small protozoa, and, sometimes, other nematodes. Some have teeth, and some have a spear that they can insert into their prey. They use the spear like a straw to suck in fluids from their prey.

Nematodes Under the Microscope

Nematodes are microscopic worms that come in a variety of sizes. These are not the worms that are sometimes found in the treatment system that you can see with the naked eye. Sometimes midge fly larvae or tubifex worms are mistaken for nematodes. These are much larger. Nematodes are not segmented, and they move throughout the fluid by wiggling their bodies back and forth (Figure 6-4).

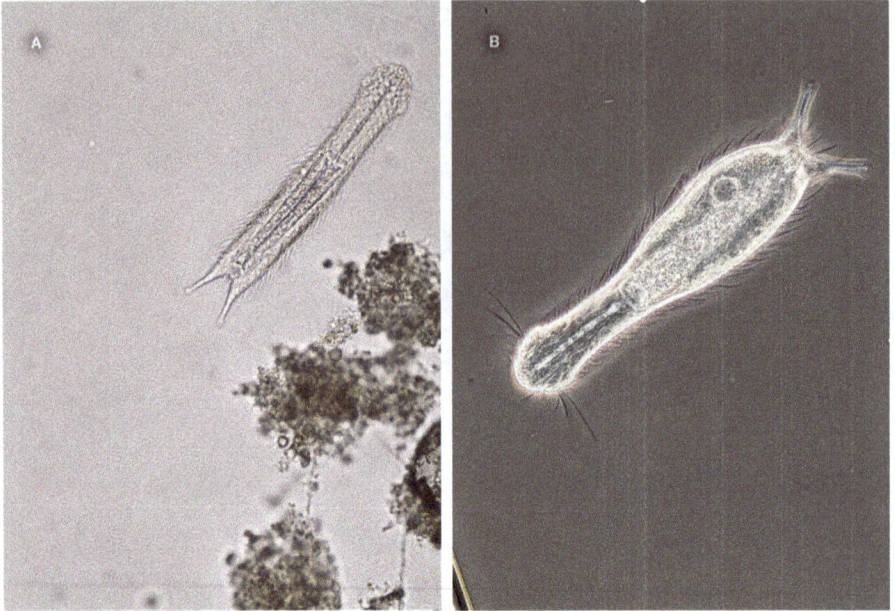

Figure 6-3   Gastrotrichs (40× magnification; phase contrast) (both A and B)

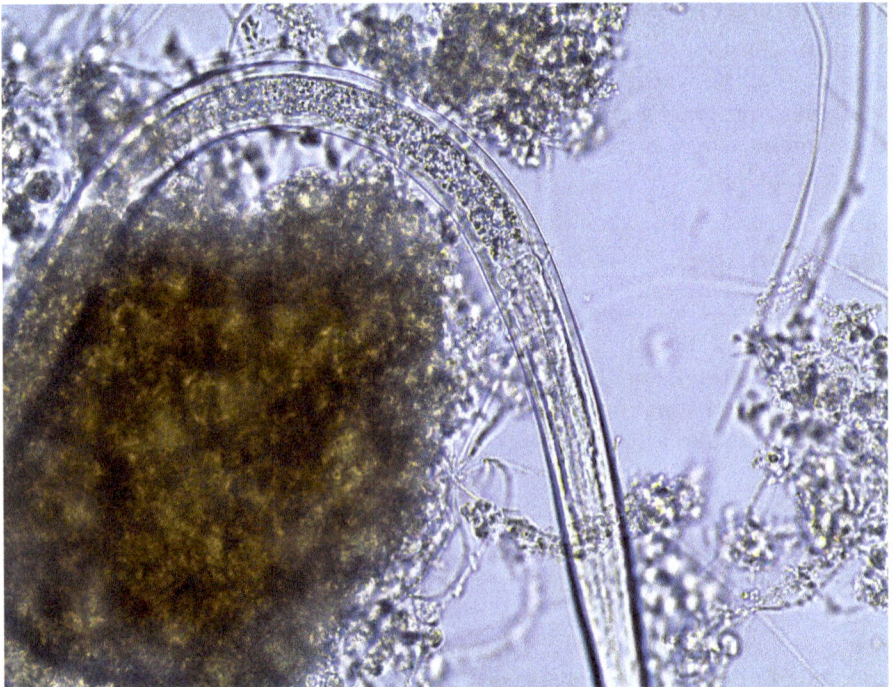

Figure 6-4   Nematode (40× magnification; phase contrast)

TARDIGRADES (WATER BEARS)

The word "tardigrade" means slow walker. They are aptly called water bears, which describes their slow, clumsy movement (Figure 6-5). They are aquatic microorganisms that depend on water to find food, breathe, reproduce, and move. They are commonly found in the same environment with rotifers and nematodes. They have developed ways to survive extreme environmental swings through a process called *cryptobiosis*. Cryptobiosis is a state of extreme inactivity in response to adverse environmental conditions. During this state, water bears lose up to 97% of their body moisture and shrivel up into a small husk called a tun (Robertson 2022). This process enables them to survive without water. Also, when there is a lack of oxygen, they can swell up and float for a few days (Miller 1997). Although they can survive extreme environmental swings, they are sensitive to ammonia and toxic conditions. Despite being associated with a longer sludge age, water bears in the wastewater treatment system can also indicate that the system is efficiently removing ammonia or that other toxic substances are not present.

Water Bears Under the Microscope

Water bears have five body segments and four pairs of short, stumpy legs with claws. They can have red, orange, or green bodies. Their head has eyes with a mouth that is used to pierce their food before sucking out the inner parts. They feed on algae and small protozoa, metazoa, and even other

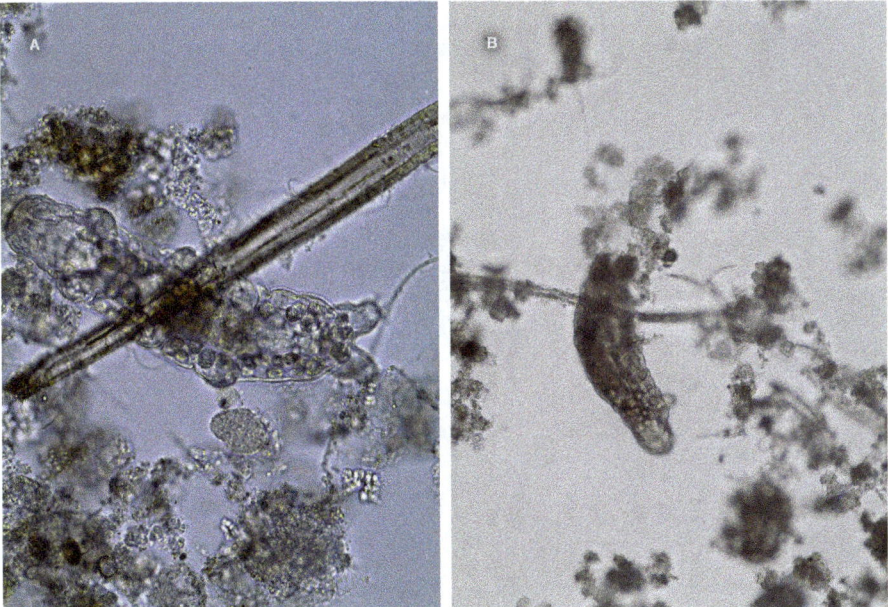

Figure 6-5   Tardigrade (water bear) at 40× magnification (A) and at 20x magnification (B) (phase contrast)

water bears. In some species of water bears, the female will molt out of her skin (termed *cuticle*) and lay her eggs inside the shed cuticle. The male will deposit his sperm inside the shed cuticle, and sometimes, he will deposit his sperm before the cuticle is shed. The eggs covered with sperm can hatch in as little as 14 days (Figure 6-6). Water bears can molt up to 12 times.

BRISTLE WORMS

The worms we generally call bristle worms in the activated sludge system are from the class Polychaeta and in the *Aeolosoma* family (Figure 6-7). They are scavengers that feed on algae and any digestible solid waste found in the sludge and are often called suction-feeding worms. They feed by placing themselves over the food, creating a vacuum.

Bristle Worms Under the Microscope

Bristle worms are usually found where there is a lot of sludge. They often embed and anchor themselves to the sludge. The bodies of these worms are often seen with brightly colored glands that look like dots, which can be red, green, blue-green, yellow, and sometimes colorless. They have transparent bodies that can be divided into 10 or more segments.

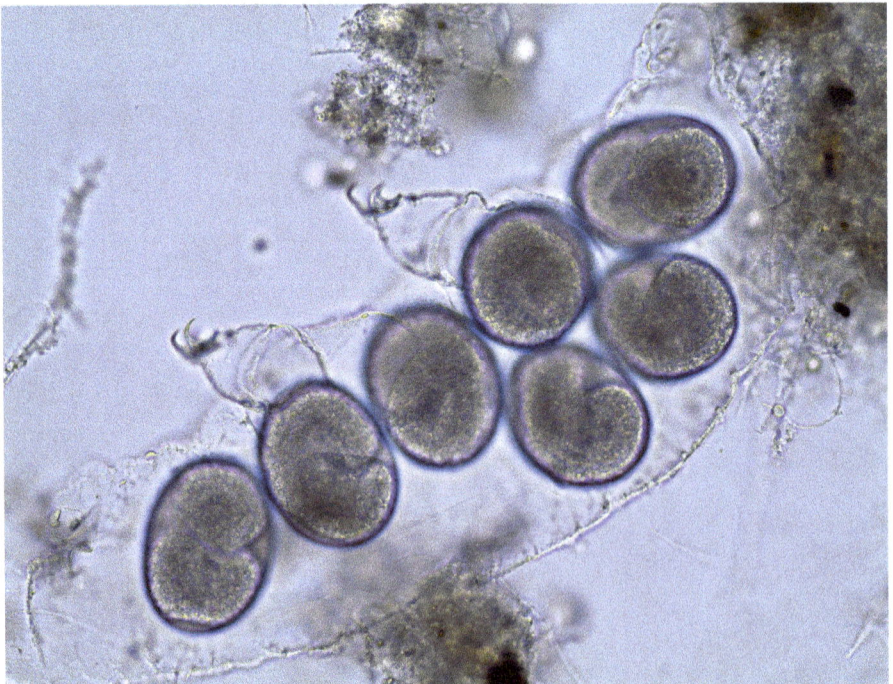

Figure 6-6　Tardigrade eggs deposited in a shed cuticle (40× magnification; phase contrast)

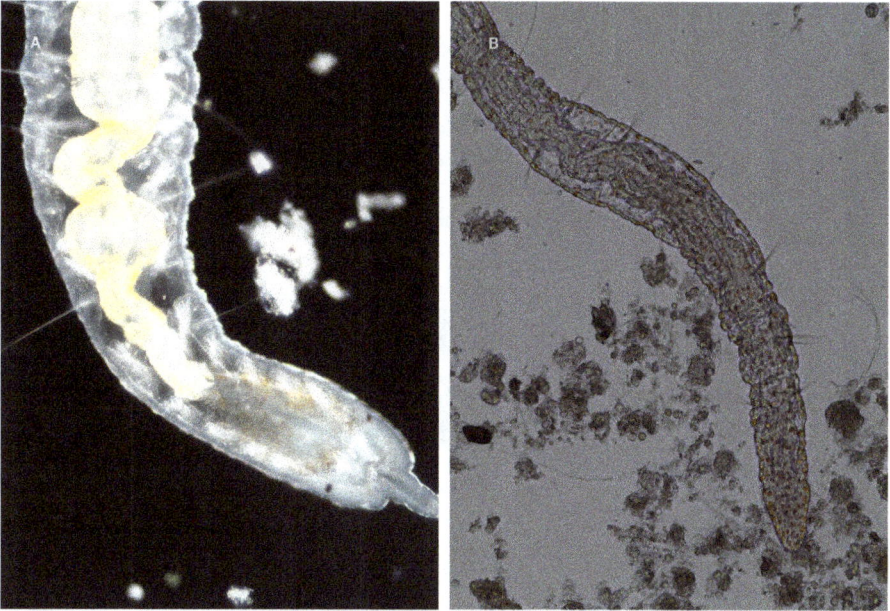

Figure 6-7   Bristle worm (darkfield illumination) (A); bristle worm (20x magnification; phase contrast) (B)

METAZOA IN ACTIVATED SLUDGE

Protozoa, bacteria, and single-celled microorganisms can multiply by dividing into two identical cells; however, metazoa are multicellular and include male and female species that produce sperm and eggs. However, they can reproduce by both sexual and asexual reproduction. Because metazoans are multicellular and have more complex body systems, they require more time to multiply. Remember, bacteria can double in number in 20 min and protozoa in 5–20 hours. Metazoa, on the other hand, can take three to four days to multiply. Because they require more time to multiply, the presence of significant numbers of metazoans suggests an older sludge age.

REFERENCES

Glymph-Martin, T. 2024. *Activated Sludge Microbes Poster 1*. Wastewater Microbiology Solutions.

Miller, W.R. 1997. "Tardigrades: Bears of the Moss." *The Kansas School Naturalist*. 43(3).

Robertson, L. 2022. "Everything You Need (and Want) to Know About Tardigrades." *Front Line Genomics*, Oct. 18. https://frontlinegenomics.com/everything-you-need-and-want-to-know-about-tardigrades/ (accessed August 16, 2024).

Chapter 7

Filamentous Bacteria and Fungi

Under normal conditions in the activated sludge treatment system, bacteria occur singly, in pairs, in small groupings, or in short chains. However, when operational conditions drastically change, the bacteria that grow in long filaments begin to gain an advantage. Changes in temperature, pH, dissolved oxygen (DO), sludge age, or even the amounts of available nutrients such as nitrogen, phosphorus, oils, and grease can affect these bacteria. The dominance of filamentous bacteria in the activated sludge treatment system can cause problems with sludge settling. At times, excessive numbers of filamentous microorganisms interfere with floc settling, and the sludge becomes bulky. This bulking sludge settles poorly and can leave behind a turbid effluent. Some filamentous microorganisms can cause foaming in the aeration basin and clarifiers.

Filamentous bacteria often live in the system in a nonfilamentous growth phase. In this phase, they are harmless and contribute to the overall removal of biochemical oxygen demand (BOD). Some filamentous bacteria are always present, even in a well-operating system. The presence of some filamentous bacteria is helpful. The filaments can provide a support structure for bacteria to attach to as they form floc. However, when they gain dominance and begin to extend beyond the floc, they may cause problems with sludge settling and cause foaming in the treatment system.

IDENTIFYING FILAMENTOUS BACTERIA

In the past, the identification of the different types of filamentous microorganisms typically involved the use of taxonomic keys. These keys led you through a series of "yes" or "no" options that direct you to the possible causative filament. The most widely used keys were the ones prepared by Eikelboom (1977) and Jenkins et al. (1993). These keys helped positively identify the filaments. In fact, Eikelboom was the one who observed many of the filamentous organisms listed as "Types." Typically, when using these keys to identify filamentous bacteria, one would look for several of the following characteristics:

- *Filament shape and size.* Filamentous bacteria grow in a variety of sizes and shapes. Filaments can be curved, straight, or irregularly shaped. They also come in different lengths. A microscope eyepiece with a measuring grid can be used to measure the length and width of the cell or filament.
- *Cell shape.* Because filamentous bacteria are actually "chains" of bacteria, the individual cells of the filament are made up of bacteria of different shapes and sizes. Some filaments are made up of round bacteria; others are rectangular, square, discoid, oval, rod, or even barrel shaped.
- *Cell septa (with or without indentation).* Septa are the cross walls that divide the filament into individual cells. Some filaments appear to be made up of one long cell, and septa are not visible. Some of the cells may even indent at the septa; others do not.
- *Sheath.* The sheath is a nonliving rigid structure that covers the filament. It resembles a clear soda straw with individual cells stacked inside (Figure 7-1).
- *Branching.* Some filaments exhibit true or false branching. False branching is caused by the physical attachment of one or more branches to the main filament. There is no continuum between the branches. True branching is actually the result of the emergence of new cells with continuum between branches (Figure 7-2).

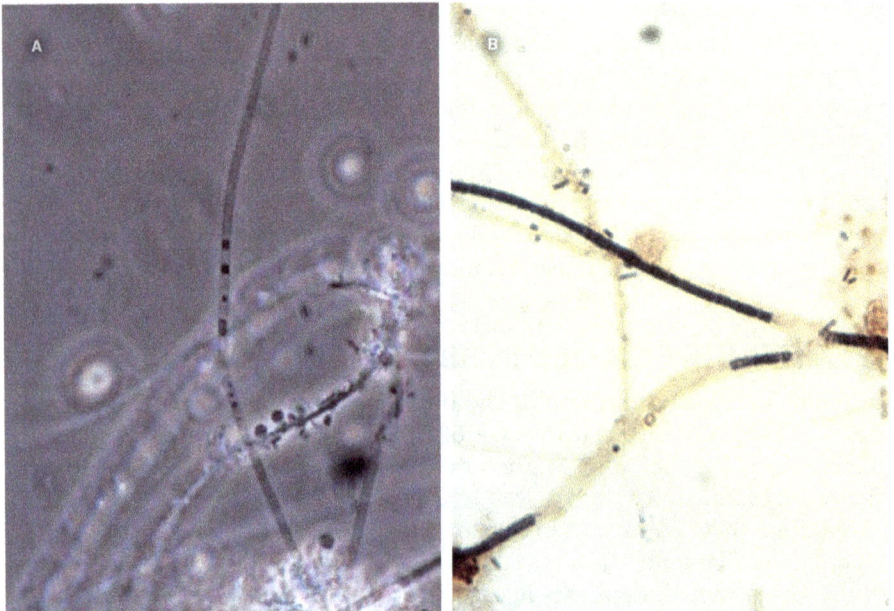

Figure 7-1   Filament sheath (100x magnification; oil immersion) (A). Filament sheath (100x magnification, oil immersion, brightfield; Gram stain) (B)

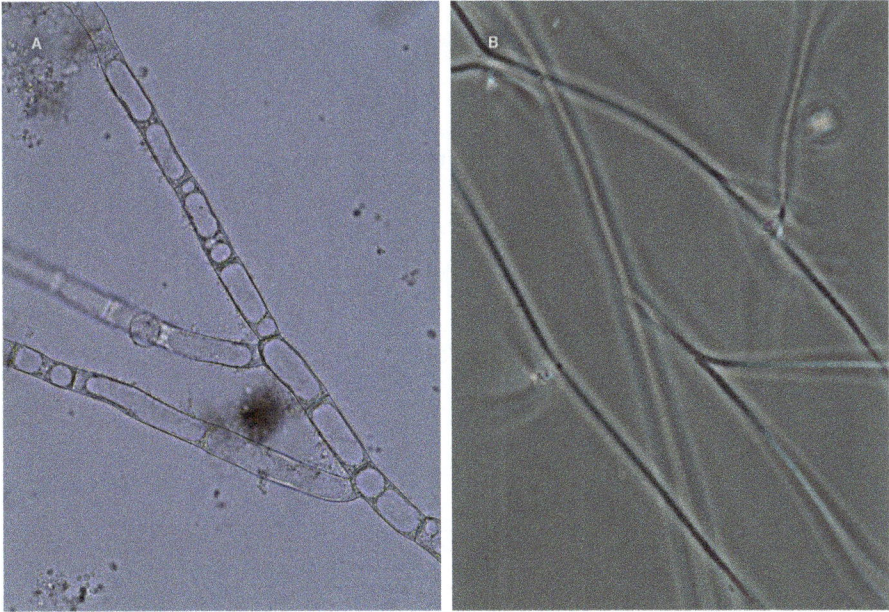

Source: Image B printed with permission from Glymph-Martin 2024

Figure 7-2 True branching (A) and false branching (B) (100× magnification; oil immersion; phase contrast)

- *Sulfur granules.* Some filaments can accumulate sulfur granules inside their cells. When viewed under the microscope with phase contrast, the granules look like light spots within the cells of the filament (Figure 7-3).
- *Epiphyte (attached growth).* Another characteristic that is common only to sheathed filaments is the presence of single-celled bacteria that grow perpendicular along the outer surface of the filamentous organism. This attached growth is called epiphyte (Figure 7-4).
- *Motility.* Motility, the filament's ability to swim, is probably the easiest characteristic to identify.

Recognizing these filament characteristics can be helpful. However, looking for all these characteristics and measuring lengths and widths of cells and filaments can be very time-consuming and requires a lot of skill. Let's look at a way to make this identification process a bit easier.

Keeping It Simple

There are two problems caused by filamentous bacteria: sludge foaming and sludge bulking. Generally, the filaments that cause bulking do not cause foaming, and vice versa. There is one filament that does both, but we will discuss that later. The first step is to determine whether the treatment system is experiencing issues with foaming or whether the problem is with bulking sludge.

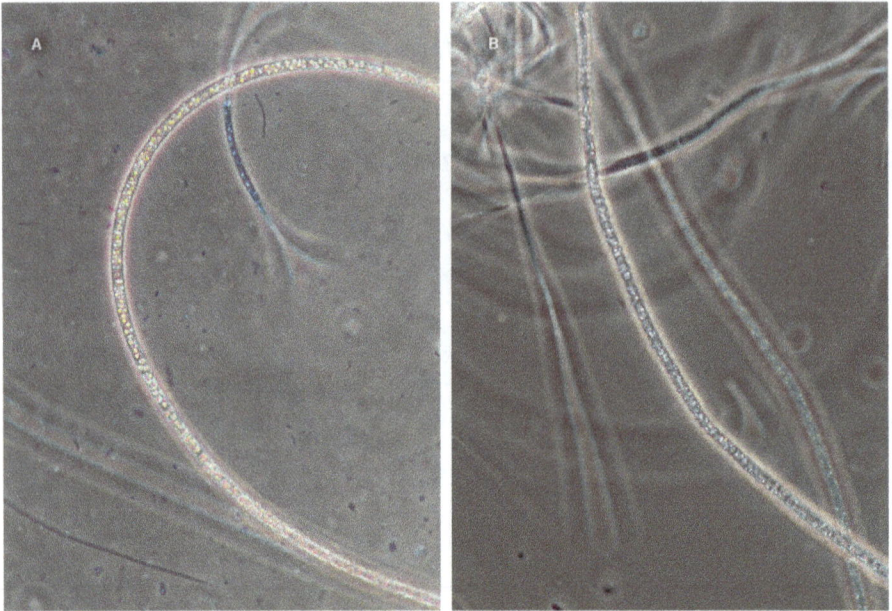

Figure 7-3 Beggiatoa and Thiothrix Type II (A and B, respectively; both 100× magnification; oil immersion; phase contrast)

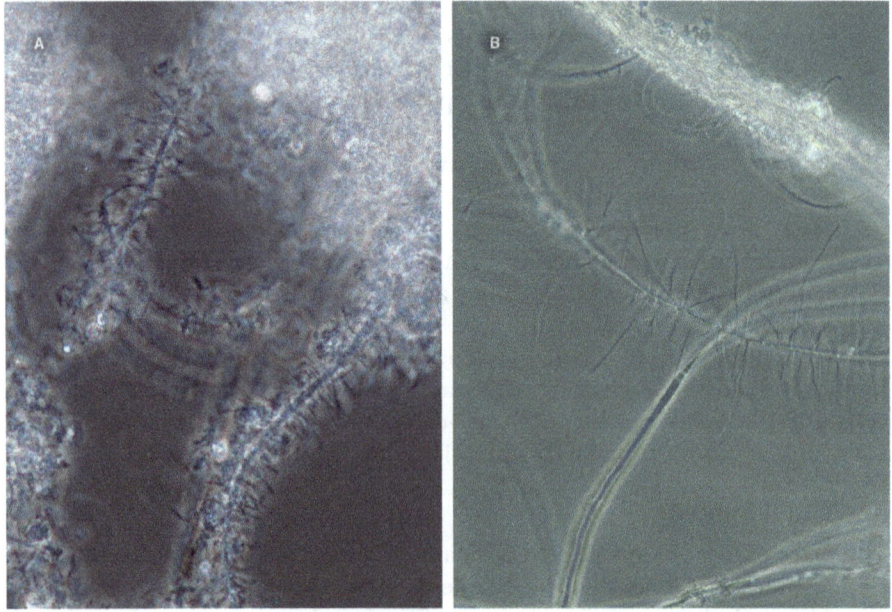

Figure 7-4 Type 0041 and Type 1701 (A and B, respectively; both 100x magnification; oil immersion; phase contrast)

Identifying foaming filaments is relatively straightforward because only three common filaments cause foaming. Two are gram-positive, and the other is gram-negative. The two gram-positive filaments are structured so differently that it is easy to tell them apart. The same is true for the gram-negative filament that causes foaming. Bulking filaments, on the other hand, require a few more steps. Observing the filaments live with the phase contrast condenser and the oil-immersion lens, and grouping the filaments based on certain characteristics, will help to narrow the field. For instance, seven of the common filaments have sheaths. Four of those with sheaths have attached growth. Four filaments are often seen with sulfur granules within the cell, two branch, and one is motile.

First, the dominance of filamentous bacteria as the cause of the foaming or bulking must be confirmed. There are other issues that can cause sludge bulking and foaming, so you want to make sure that filamentous bacteria are the cause. This can be determined by making a wet mount and examining a sample of mixed liquor under the microscope with the 20× or 40× objective. Or you can make a smear, wait for it to dry, and stain it using the Gram stain. If filamentous bacteria are causing the problem, their dominance will be apparent in the sample. If the treatment system is foaming, the filaments will be in the foam. If the issue is sludge bulking, the filaments will be observed in the floc and extended into the bulk fluid. Generally, the Gram stain can help to identify most of the common filamentous bacteria. However, the wet mount should be used to view the characteristics mentioned earlier in this chapter. Let's look at filaments that cause foaming first.

All filamentous bacteria should be observed using phase contrast and oil-immersion 100× objective. Stained slides should be observed using oil-immersion 100× objective under bright-field, not phase contrast.

FILAMENTOUS BACTERIA THAT CAUSE SLUDGE FOAMING

Foaming occurs when certain types of filamentous bacteria float to the surface, bringing solids and air bubbles with them. Filamentous bacteria that cause foaming have hydrophobic cell walls. The cells are covered with a waxy coating that resists water and enables them to float when aerated. They have very little effect on the settleability of the sludge or the overall treatment of the wastewater, unless they cause excessive amounts of foam. Foam can cover the surface of wet wells, aeration basins, and clarifiers and can spill over the weirs, causing solids loss, elevated effluent suspended solids, and even freezing in the winter.

If the system is foaming, simply examine a sample of the foam. If filamentous bacteria are causing foaming, the filaments will be dominant in the foam. Once it has been established that foaming is due to the excess growth of filamentous bacteria, you can begin the process of identifying certain characteristics.

Figure 7-5 Gram stain: Gram-positive (+) stains are purple; gram-negative (–) stains are pink.

There are only three common filaments that cause foaming. *Nocardia* spp., *Microthrix parvicella*, and Type 1863 are the major causes of foaming in activated sludge treatment systems. These are very easy to identify.

To identify the type of filament causing the foam, first collect a sample of foam and make a smear. Allow the smear to air dry, and then stain it with Gram stain. The Gram stain will help to narrow your choices to two: gram-positive or gram-negative. Two of the filaments that cause foam are gram-positive, and the other is gram-negative (Figure 7-5). Observe the gram-stained slide with a drop of immersion oil with the 100× objective. Purple-stained filaments are gram-positive, and pink-stained filaments are gram-negative. *Nocardia* and *M. parvicella* are both gram-positive. Type 1863 is gram-negative.

Type 1863

Type 1863 appears as irregularly shaped clusters and stains gram-negative. It often looks like a pink-dashed line when stained and is easy to identify (Figure 7-6). Just like the other foam-causing filaments, this filament has hydrophobic cell walls that resist water and enable it to float. It thrives in the presence of excess amounts of fats, oils, and grease (FOG). However, the unique conditions that favor its growth over the other foaming filaments are a sudden decline in aeration basin pH.

M. parvicella

M. parvicella also has a unique appearance. It stains gram-positive and is thin and smoothly curved. In fact, it looks like "purple spaghetti" (Figure 7-7). This filament uniquely thrives in the presence of excess amounts of animal and vegetable FOG, like those of kitchens and high-density restaurants. It also has the unique ability to thrive in colder temperatures. Most often, the foaming in the winter months is caused by this filament. *M. parvicella* can also grow rapidly in low–food-to-microorganism (F/M) ratio and low-nutrient conditions. This is also the one filament that can cause both foaming and bulking in activated sludge.

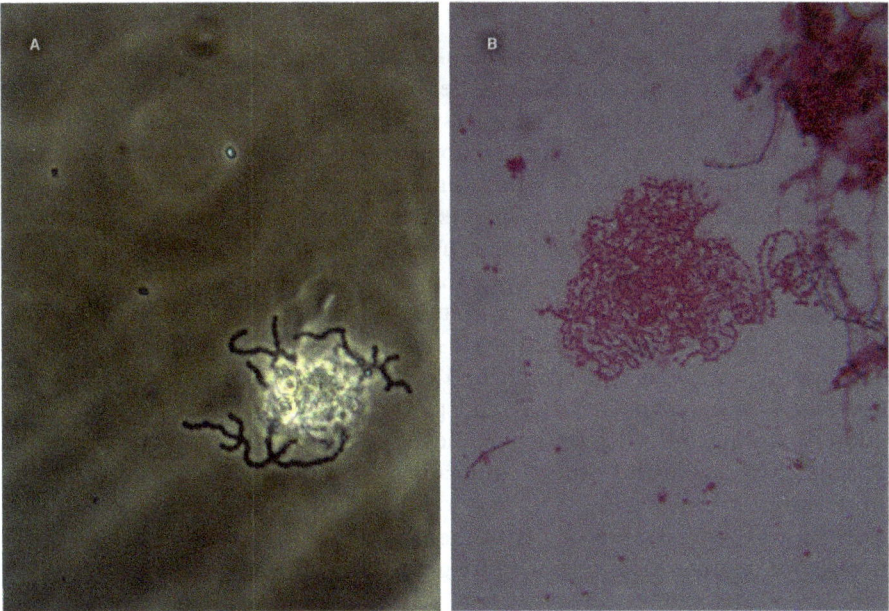

Figure 7-6 Filament Type 1863: Wet mount (100× magnification, oil immersion, phase contrast) (A) and Gram stain (100x magnification, oil immersion, bright-field) (B)

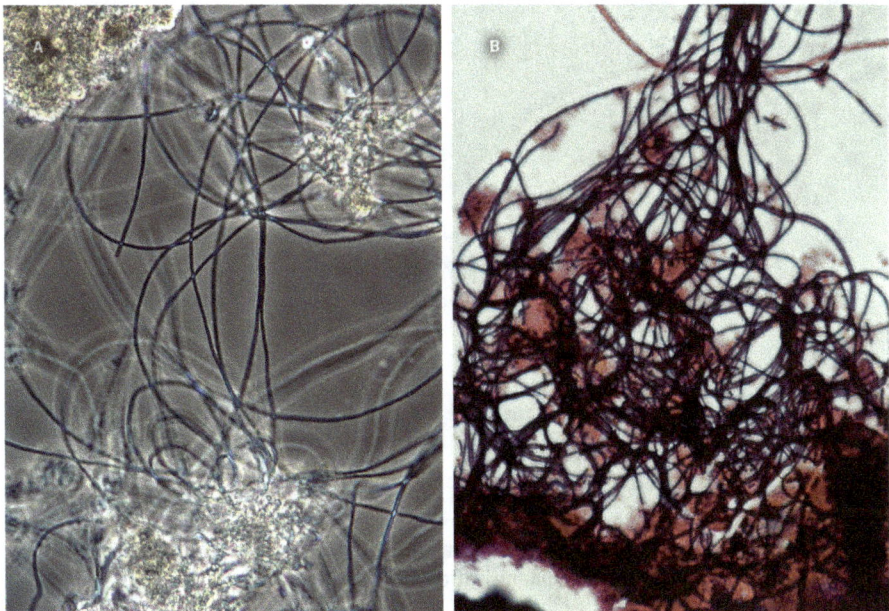

Source: Image A printed with permission from Glymph-Martin 2024

Figure 7-7 *Microthrix parvicella*: Wet mount (100× magnification, oil immersion, phase contrast) (A) and Gram stain (100× magnification, oil immersion, bright-field) (B)

Nocardia

Nocardia is the most common foam-causing filament in activated sludge. The strongly hydrophobic branches enhance sludge flotation by forming a net, enabling them to trap oil droplets and gas bubbles. *Nocardia* likely produces surface-active substances in abundance, altering the surface tension and causing foaming when aerated. Under certain conditions, *Nocardia* produces a lipid material, which, when excreted into the mixed liquor, collects on the surface of air bubbles. This causes the bubbles and *Nocardia* to mesh together, giving the foam the appearance of chocolate parfait. *Nocardia* is a gram-positive filament that exhibits true branching. It looks like small patches of irregularly branched short filaments (Figure 7-8).

Upon aeration, it is the branched cluster (adult stage) that forms a net, enabling the entrapment of oil and gas bubbles, causing foam upon aeration. Foaming only occurs when the population of adult stage *Nocardia* is quite high. Like the other foam-causing filaments, *Nocardia* is favored in the presence of excess FOG and low F/M ratio; and because it is a very slow grower, when compared with the traditional floc-forming bacteria, it is favored in treatment systems with longer mean cell residence times (MCRTs).

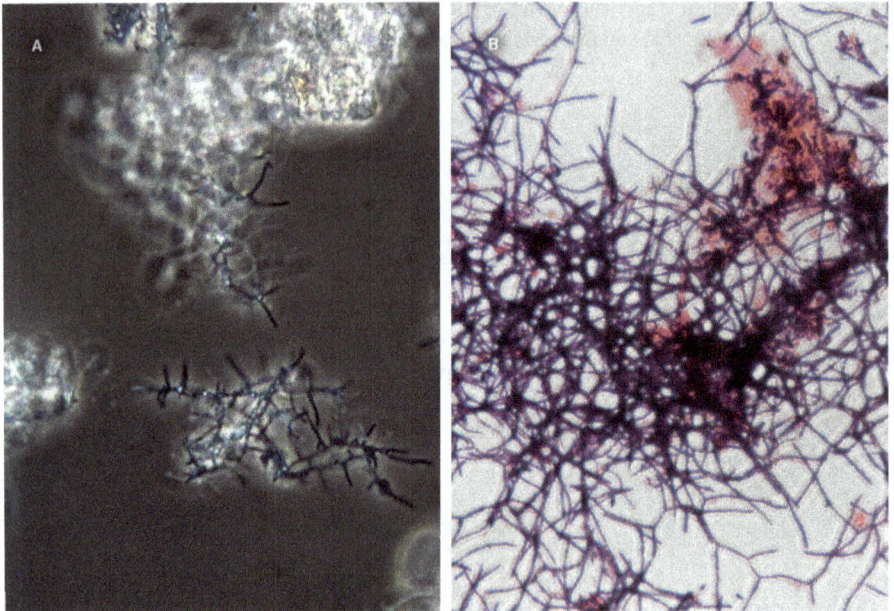

Figure 7-8 *Nocardia* sp.: Wet mount (100× magnification, oil immersion, phase contrast) (A) and Gram stain (100× magnification, oil immersion, bright-field) (B)

CONTROLLING FOAMING FILAMENTOUS BACTERIA

Filamentous bacteria that cause foaming are all associated with low–F/M ratio conditions and excess FOG. They all have hydrophobic cell walls that enable them to float and create a stable foam. *Microthrix* can be controlled by increasing the F/M ratio and controlling greases and oils entering the aeration basin. Primary treatment plays a critical role in removing grease and oils. During primary treatment, greases and oils rise to the surface and are skimmed off and removed for further treatment. In treatment systems without primary clarification, grease and oils pass directly into the aeration basin, creating favorable conditions for foam-causing filamentous bacteria.

Filament Type 1863 can also be controlled by increasing the F/M ratio and minimizing grease and oils entering the aeration basin. However, this filament is favored whenever there is a significant decline in the aeration basin pH. So, in addition to controlling grease and oils, maintaining a stable pH in the range of 6.5–8.5 is recommended to control the growth of this filament.

Nocardia, on the other hand, is a little more difficult to control. What makes this filament so difficult? *Nocardia* is uniquely different from the other foam-causing filaments. It has a distinctive cell wall. The cell wall is viscous and contains a high concentration of lipids (fats). The cell walls are also hydrophobic (resist water) and can float easily to the water surface. *Nocardia* is one of the most difficult filaments to eliminate from the treatment system. It competes successfully for several reasons. The hydrophobic cell walls and branched hyphae allow oil droplets and gas bubbles with accumulated nutrients on their surface to stick to the *Nocardia* cell walls. Also, fats, proteins, and other nutrients that float on the surface of the water accumulate and supply more nutrients than those in the bulk water. *Nocardia* can also live on dead cells and can use organic compounds adsorbed on floc that are only slowly degraded by other organisms. This allows them to survive periods when dissolved nutrients are scarce. They can also store polyphosphates as reserve material, which enables them to increase in the system even when they are virtually cut off from a nutrient supply. As a former soil inhabitant, *Nocardia* is better adapted than its competitors to dryness and ultraviolet radiation that prevail in the foam.

One reason why there is an enormous increase of *Nocardia* during foaming is its ability to switch from a "K"-growth strategy to a "μ"-growth strategy. Most bacteria are K-growth strategists—i.e., they can efficiently take up nutrients so that their growth rate is very steep when high concentrations of nutrients are available. As the level of nutrients diminishes, the growth curve levels off and then begins to decline. Those that are μ-growth strategists have a gradual growth curve but can reach much higher growth rates over time. They are also adaptable to local and seasonal fluctuations of nutrients and can form resting stages during starving periods. The ability to switch from one strategy to another allows *Nocardia* to maintain a stable background population when nutrient conditions are restricted and competition by other organisms is strong.

Nocardia is also adaptable. When substrate concentrations are high (high F/M ratio), *Nocardia* can grow almost 100 times faster than other organisms. When substrate concentrations are low, they have the ability to form resting stages within starving periods and can maintain a stable background population when nutrient conditions are restricted and competition by other organisms is strong. In addition, the foam, once accumulated, provides a source of seed for *Nocardia* growth. *Nocardia* has the ability to live on dead cells in the foam and can use leftover organic compounds adsorbed to the sludge floc, which are only slowly degraded by other organisms. Therefore, it can continue to survive and multiply in the foam.

Its growth pattern is also uniquely different. In its early growth stage, *Nocardia* is harmless in the treatment system and does not cause foaming. In this stage, it is unrecognizable as *Nocardia*. Instead of the classic branched filament, it lives as small, cylindrical, short, gram-positive (purple) cells and can go undetected. As the temperature of the wastewater begins to rise, the small cells begin to form small nodes as the branches begin to form. In this stage, *Nocardia* still does not cause any significant foaming. Once the wastewater temperature rises above 16°C (60°F), the branches begin to elongate rapidly. Not until *Nocardia*, fully branched, reaches this advanced growth stage does it cause foaming when aerated (Figure 7-9).

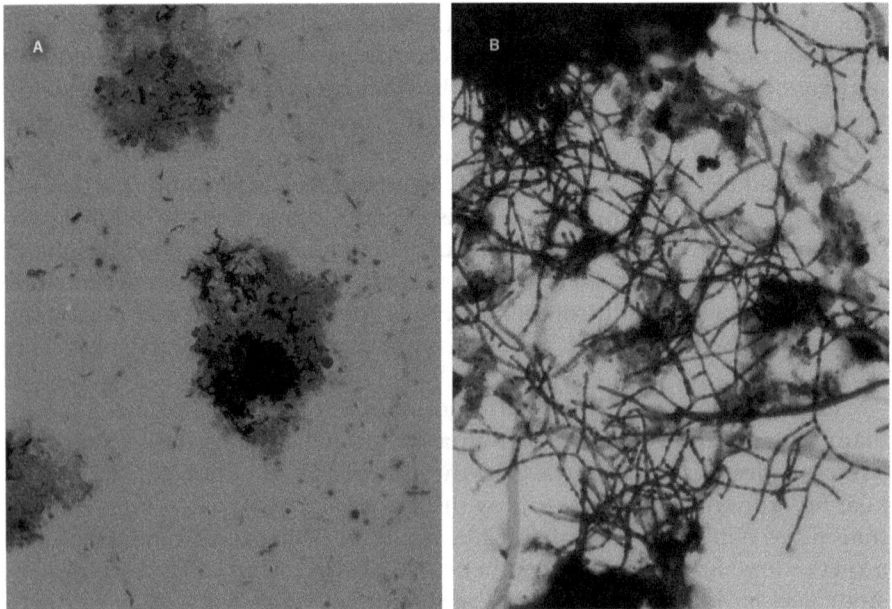

Figure 7-9 *Nocardia* sp.: Early stage (small rods and nodes) (A) and adult stage (fully branched) (B) (both Gram-stained, 100× magnification, oil immersion, bright-field)

What often happens with *Nocardia* foam is the treatment plant operator will set up chlorine sprays to spray the foam. Most likely, the foam will begin to disappear, but the problem has not been solved. A unique characteristic of *Nocardia* is that, as a part of its natural growth cycle, the branches break up into cylindrical, short cells, and the foam will begin to dissipate.

The short cells begin to grow and branch once again, and the growth cycle starts all over again. Consider this: The foam dissipates, but the new cell "babies" increase in number. They settle into the mixed liquor and begin to form nodes again, which eventually turn into branches. The increased number of branches will create a greater amount of foam during the next cycle. An operator who thinks spraying chlorine was successful in controlling the foam may increase the chlorine dosage. However, the growth cycles continue, and each generates more foam. The foaming ends up even worse!

Why?

Chlorine works well for most filamentous bacteria because the filaments have much more surface area than the floc-forming bacteria. Chlorine can affect the filaments without adversely affecting the other microorganisms. But this is not so with *Nocardia*. Applying chlorine sprays to the foam will cause the *Nocardia* branches to prematurely break up into tiny cells and decrease the surface area. Therefore, in essence, the operator is delivering "premature babies" that will eventually turn into branched filaments that cause more foam.

So, What Should the Operator Do?

Floc-forming bacteria grow rapidly in the activated sludge system, peaking in number early in the process. *Nocardia*, however, requires a much longer time to grow from its early growth stage to fully branched filaments. To a great degree, the sludge age determines whether *Nocardia* will remain in the early growth stage and cause no foam or will mature to fully branched filaments and cause foaming in the system. As long as the sludge age remains short enough, *Nocardia* will continue to live as harmless, small cells. *Nocardia* is generally dormant in colder temperatures but grows quickly in warmer temperatures. Once this filament gains dominance in the treatment system, it forms a stable foam that is difficult to control.

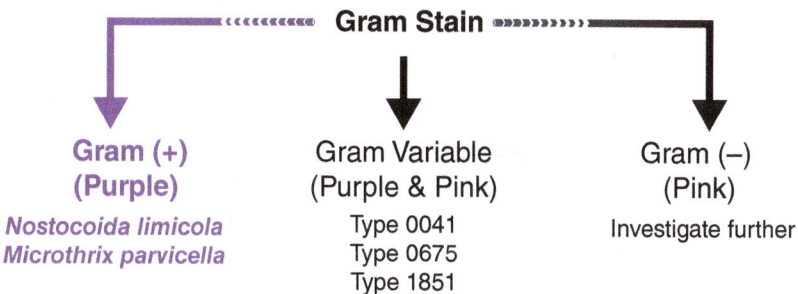

Gram (+) (Purple)	Gram Variable (Purple & Pink)	Gram (−) (Pink)
Nostocoida limicola *Microthrix parvicella*	Type 0041 Type 0675 Type 1851	Investigate further

Figure 7-10 Gram stain: Gram-positive filamentous bacteria stain purple.

What About All the Foam on the Tanks, in the Wet Wells, and on the Clarifiers?

As long as the foam remains on the surface of the aeration basin, clarifiers, and wet wells, *Nocardia* will continue its growth cycle. The foam, once accumulated, also provides a source of seed for Nocardial growth. Therefore, the foam must be sucked or skimmed off to prevent *Nocardia* from multiplying. Most importantly, the skimmed-off foam should not be put back into the head of the treatment system but should be land-applied or digested. Eliminating foaming problems caused by *Nocardia* requires time, patience, and work. By decreasing the amount of time *Nocardia* remains in the aeration basin, you prevent it from maturing into branching filaments, and removing the foam from the system will prevent it from multiplying. After one or two (or three) summers, your system should be foam-free!

FILAMENTOUS BACTERIA THAT CAUSE SLUDGE BULKING

Sludge bulking occurs when something interferes with the separation of the solids from the treated water. Remember, the key to good treatment is the separation of the treated water from the biological floc (sludge). Excessive amounts of filamentous bacteria can create bridging between floc particles, making it difficult for the solids to settle and separate from the liquid.

Table 7-2 contains a list of the most common filamentous bacteria that can cause bulking in activated sludge.

Begin by collecting a sample of mixed liquor from the discharge end of the aeration basin. Make a smear and stain it with the Gram stain. Using the Gram stain, bulking filaments will stain one of three ways: gram-positive, gram-variable, or gram-negative. There are only two gram-positive (stain purple) filaments: *Nostocoida limicola* and *M. parvicella*.

With the exception that both stain positive (purple) with the Gram stain, *N. limicola and M. parvicella* are very different from each other and are easy to identify. *N. limicola* has three types: I, II, and III. Structurally, they are slightly different. *N. limicola* Type I is the smallest of the three. It is irregularly shaped and has oval cells, and the cell septa are hard to see. Type II, slightly larger, is also irregularly shaped with oval-shaped cells, but the cell septa are clearly seen. Type III is the largest and has round- to oval-shaped cells with very clear indentations. All three types are irregularly shaped with oval to round cells, and all three stain gram-positive (although Type II can stain gram-negative at times). *Nostocoida* often looks like a "beaded necklace or string of pearls" (Figure 7-11). The growth of this filament is associated with low–F/M ratio and sometimes low-DO conditions and with waste containing starch. It is more often seen in significant amounts in industrial treatment facilities.

Table 7-1 Filamentous bacteria that cause sludge bulking

Nostocoida limicola	Type 0041
Sphaerotilus natans	Type 0675
Haliscomenobacter hydrossis	Type 1851
Thiothrix I & II	Type 0092
Microthrix parvicella	Type 1701
Beggiatoa	Type 021N
Type 0961	Type 0914

Table 7-2 Gram-positive filamentous bacteria that cause sludge bulking

Filament Name	Conditions That Can Favor Their Growth
Nostocoida limicola Types I, II, and III	• Low DO • Wastes high in starch
Microthrix parvicella	• Animal and vegetable FOG • Low F/M ratio • Colder temperatures

DO—dissolved oxygen, F/M ratio—food-to-microorganism ratio, FOG—fats, oils, and grease

Table 7-3 Gram-variable filamentous bacteria that cause sludge bulking

Filament Name	Appearance	Conditions That Can Favor Their Growth
Type 0041	Large filament with heavy attached growth.	• Low F/M ratio • Low nutrients
Type 0675	Slightly smaller with moderate attached growth.	• Low F/M ratio • Low nutrients
Type 1851	Thin with sparse attached growth. Often seen growing in bundles.	• Low F/M ratio • Low nutrients

On the other hand, *M. parvicella* is smoothly curved, its cell septa are difficult to see, and it looks like "purple spaghetti" when stained (Figure 7-7). *M. parvicella* is the only filament that can cause both bulking and foaming. These two filaments are distinctly different from one another in appearance, so, if you see a gram-positive filament dominating in the sample, it is either *N. limicola* or *M. parvicella*. One is irregularly shaped with oval to round cells, and the other looks like spaghetti.

Foaming filaments are not difficult to identify because there are only three of them. The process of identifying bulking filaments, on the other hand, requires a bit more investigation. The gram or Neisser stain is helpful but provides only limited information. Most bulking filaments are gram-negative

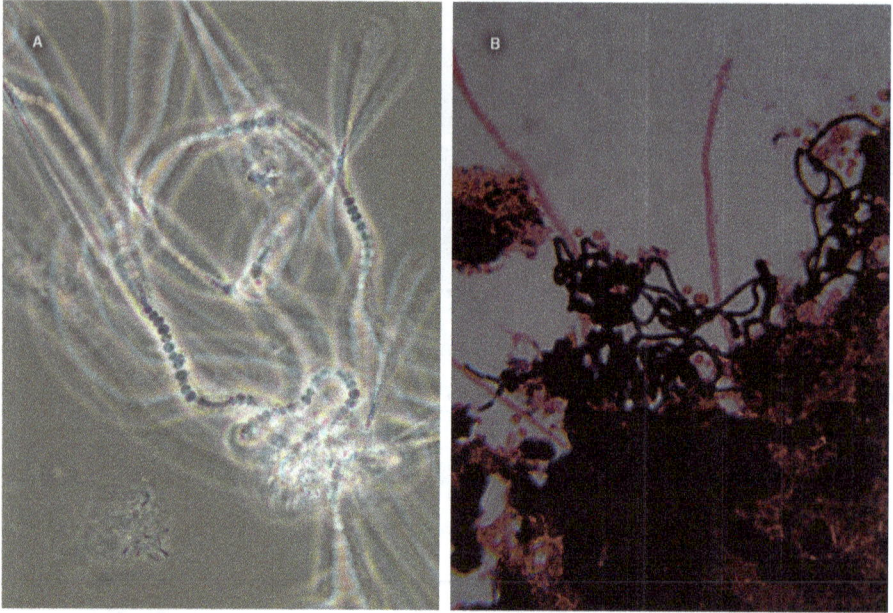

Figure 7-11 *Nostocoida limicola*: Wet mount (100× magnification, phase contrast) (A) and Gram-stained (100× magnification, oil immersion; bright-field) (B)

and Neisser-negative. A few filaments are gram-variable, meaning that they have both gram-positive and gram-negative characteristics. Generally, this means that the sheath stains gram-negative (pink), whereas the cells inside stain gram-positive (Figure 7-12).

If the filament stains gram-positive, it will be easy to identify, but what if it stains gram-variable (Figure 7-13)? If the filament is gram-variable, you have narrowed your choices to three. There are three filaments that cause bulking that stain gram-variable: Type 0041, Type 0675, and Type 1851.

Type 0041, Type 0675, and Type 1851

Filament Types 0041, 0675, and 1851 are a unique group of filaments. All have a sheath and are covered with epiphyte (attached growth). Type 0041 is the largest, has square-shaped cells, and is covered with heavy attached growth. Type 0675 is a little smaller, also has square-shaped cells, and is covered with moderate attached growth (Figure 7-14).

Type 1851 is gram-variable just like Types 0041 and 0675, but it is a significantly thinner filament, and although it also has attached growth, the growth is sparse. One unique thing about Type 1851 is that it often is observed growing in bundles (Figure 7-15).

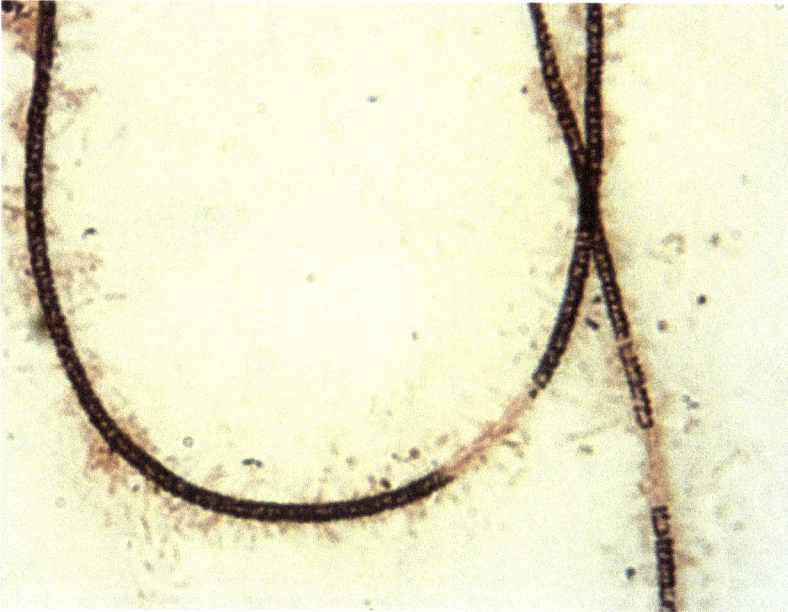

Figure 7-12 Gram-variable; gram-negative sheath and gram-positive cells within the sheath (Gram stain, 100× magnification, oil immersion; bright-field)

Figure 7-13 Gram-variable: Filament sheath stains gram-negative, and cells within the sheath stain gram-positive.

All three of these filaments are gram-variable and are covered with attached growth. They also grow under the same conditions. Their growth is favored in low–F/M ratio and low-nutrient conditions.

The good news is that you do not have to work extra hard trying to determine whether the specific filament is Type 0041, Type 0675, or Type 1851 because all three thrive under the same conditions. If you observe a gram-variable filament with attached growth, the cause is the same: low F/M ratio and low nutrients.

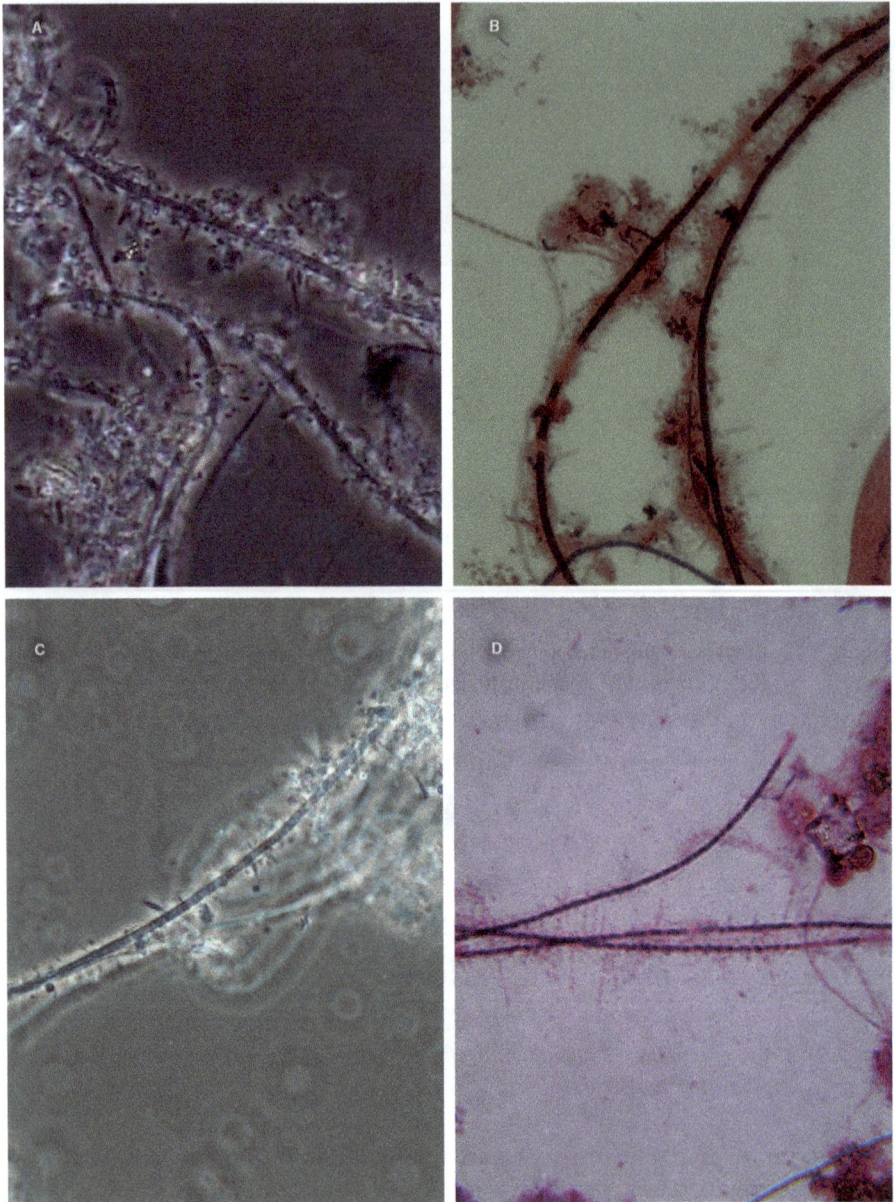

Figure 7-14 **Type 0041 wet mount (100× magnification, oil immersion, phase contrast) (A); Type 0041 Gram stain (100× magnification, oil immersion, bright-field) (B); Type 0675 wet mount (100× magnification, oil immersion, phase contrast) (C); and Type 0675 Gram stain (100× magnification, oil immersion, bright-field) (D)**

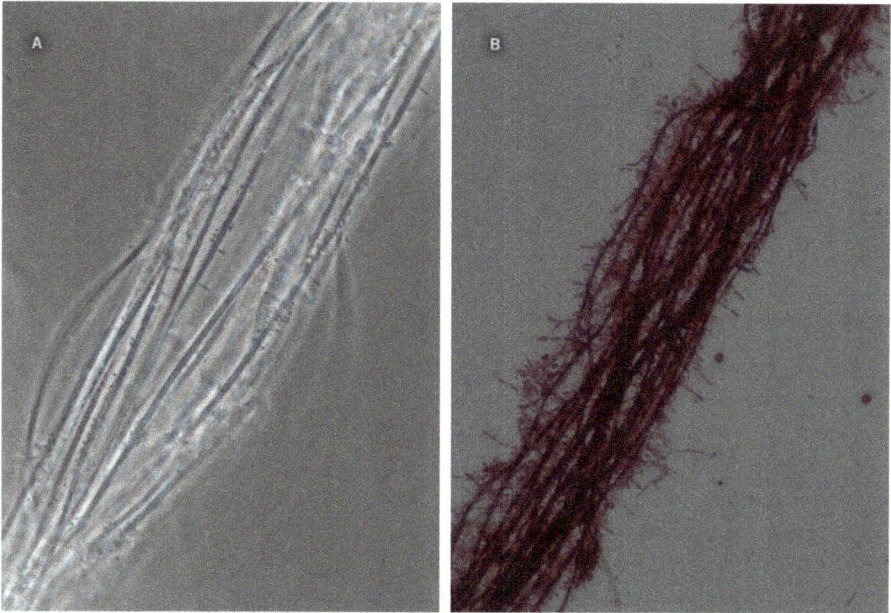

Figure 7-15 Type 1851: Wet mount (100× magnification, oil immersion, phase contrast) (A) and Gram-stained (100× magnification, oil immersion, bright-field) (B)

Figure 7-16 Gram-negative: A significant number of filamentous bacteria stain gram-negative; identifying the filament will require further investigation.

If the filament causing bulking in the treatment system is either gram-positive or gram-variable, identification is relatively easy. We know that, if it's gram-positive, it is *N. limicola* or *Microthrix,* and, if it's gram-variable, it is Type 0041, Type 0675, or Type 1851. However, most of the filaments causing bulking are gram-negative. That means that we have to investigate further (Figure 7-16).

To identify other bulking filaments, we need to first observe some of their other unique characteristics. As part of the staining procedures, the sample is dried on a slide before staining. This drying process alters the size and shape of the filament and its individual cells. So, to view these characteristics unaltered, it is best to observe filaments live with a wet mount using phase contrast with the 100× oil-immersion objective (see Chapter 3). Let's look for one or more of the following:

- Sheath
- Presence of sulfur granules
- Branching
- Motility

SHEATHED FILAMENTS

Individual cells may be stacked inside a sheath like a chain, or there may be one continuous cell inside the sheath. Cells may die or escape, leaving an empty space in the sheath. As you examine the sample, look up and down the length of the filaments for missing cells or empty spaces between cells. Of the filaments that are commonly found in activated sludge, seven of them have sheaths, as shown in Table 7-4.

Some sheathed filaments have attached growth and resemble bottle brushes. Of the common sheathed filaments, the four filaments with the "type and number" name are often seen with attached growth. This makes it easier to remember. The other three have a sheath but are not usually seen with attached growth. All of these filaments can also grow in the treatment system without the attached growth. This usually means that conditions are such that they can grow quite rapidly. This also makes them a little more difficult to identify. Identifying the sheath will help to narrow your options to these seven, but additional observations are necessary to determine which one is dominant in the treatment system.

Sheathed Filaments With Attached Growth

The filaments that have a sheath with attached growth are Types 0041, 0675, 1851, and 1701. Remember, Types 0041, 0675, and 1851 are also gram-variable filaments. See the discussion about gram-variable filaments for their

Table 7-4   Filamentous bacteria with a sheath

Sheath With Attached Growth	Sheath Without Attached Growth
Type 0041	Thiothrix I and II
Type 0675	Sphaerotilus natans
Type 1851	Haliscomenobacter hydrossis
Type 1701	

description and conditions that favor their growth. The remaining sheathed filament with attached growth is Type 1701.

Type 1701

Type 1701 is gram-negative and has cells that look like round-ended rods or "link sausages" that are stacked in a tight-fitting sheath. Type 1701 is a very thin filament and is often seen with heavy attached growth. The attached growth on this filament is generally much longer than on the others. However, there are times when this filament grows so rapidly in the system that growth does not have time to attach. Even with or without heavy attached growth, it is easy to identify (Figure 7-17). Type 1701 is favored in low-DO conditions.

Sheathed Filaments Without Attached Growth

Sphaerotilus natans, Thiothrix spp., and *Haliscomenobacter hydrossis* are sheathed filaments but are generally not seen with attached growth. Nevertheless, they are very different from each other and easy to identify.

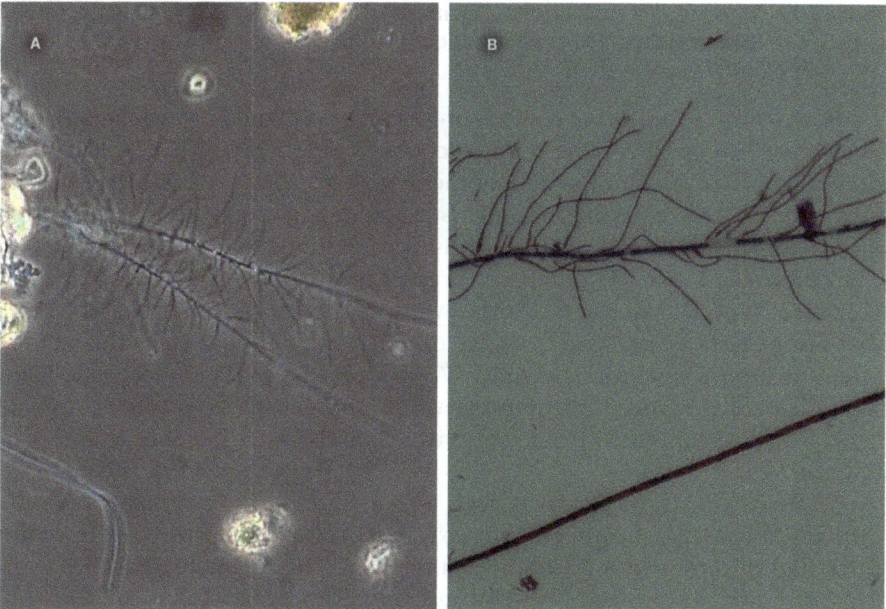

Source: Image A printed with permission from Glymph-Martin 2024

Figure 7-17 Type 1701: Wet mount (100× magnification, oil immersion, phase contrast) (A) and Gram stain (100× magnification, oil immersion, bright-field) (B)

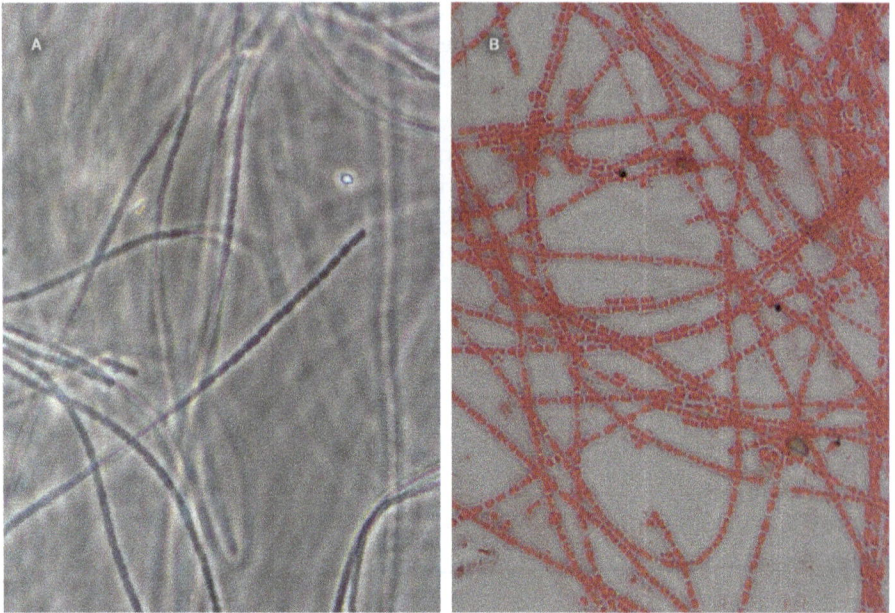

Figure 7-18 *S. natans*: Wet mount (100× magnification, oil immersion, phase contrast) (A) and Gram-negative (100× magnification, oil immersion, bright-field) (B)

S. natans. *S. natans* is also gram-negative and has sausage-shaped cells like Type 1701, but it is more than twice the width of Type 1701. It is also the only filament that exhibits false branching. Also, like Type 1701, *S. natans* is favored in low-DO conditions. This filament grows easily in areas where there is oxygen stress or dead zones in the aeration basin (Figure 7-18).

Thiothrix **I and II.** Thiothrix Types I and II have barrel-shaped cells. However, Type II is about twice the width of Type I. Another characteristic that distinguishes *Thiothrix* from the other sheathed filaments is the storage of sulfur granules within the cells. We will discuss that characteristic later. This filament is favored in the presence of appreciable amounts of organic acids, sulfur compounds from septic waste, and waste deficient in nitrogen (Figure 7-19).

H. hydrossis. H. hydrossis is such a thin, straight filament that it can be difficult to detect a sheath. *H. hydrossis* most commonly protrudes from the flock like thin pins in a cushion (Figure 7-20). This filament also thrives in low-DO, low-nutrient, and low–F/M ratio environments.

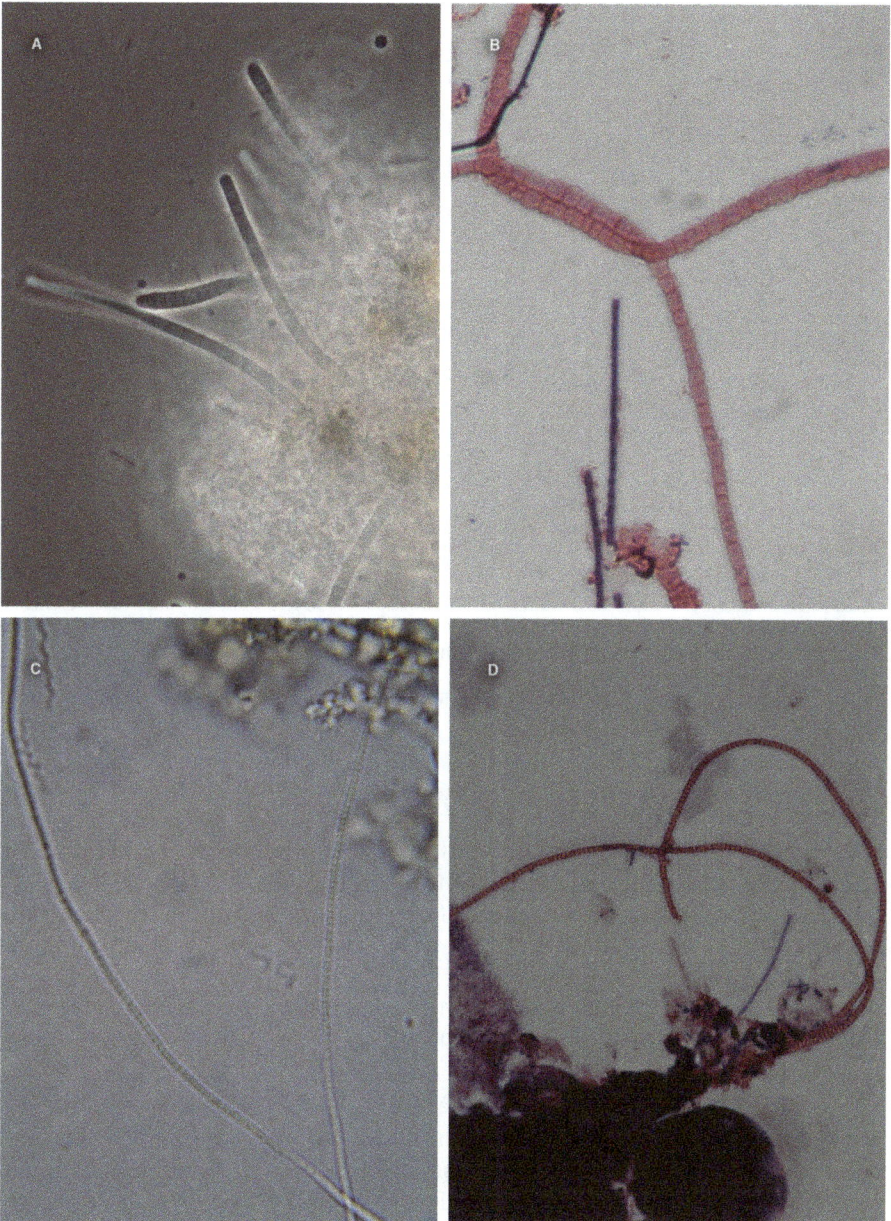

Source: Image B printed with permission from Glymph-Martin 2024

Figure 7-19 *Thiothrix* Type I wet mount (100× magnification, oil immersion, phase contrast) (A); *Thiothrix* Type I Gram stain (100× magnification, oil immersion, bright-field) (B); *Thiothrix* Type II wet mount (100× magnification, oil immersion, phase contrast) (C); and *Thiothrix* Type II Gram stain (100× magnification, oil immersion, bright-field) (D)

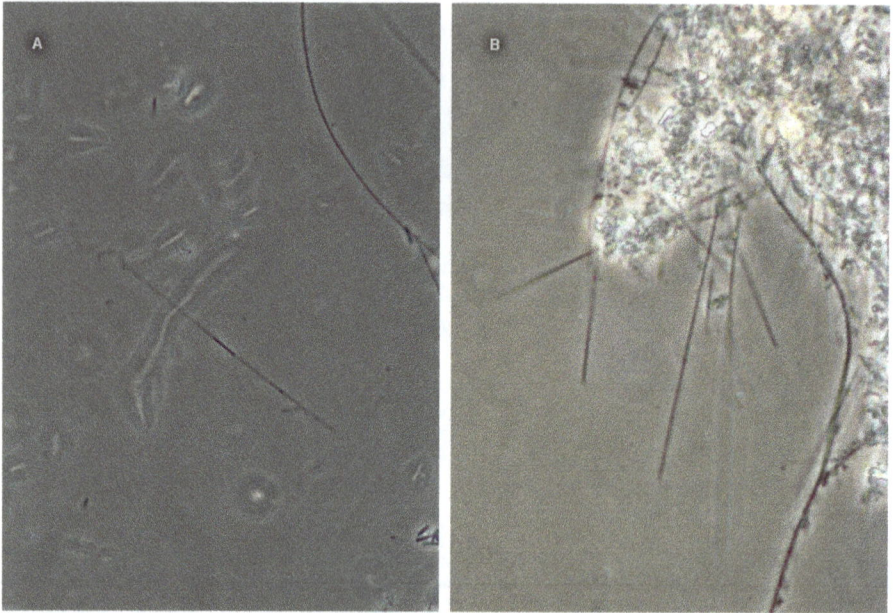

Figure 7-20 *H. hydrossis*: **Wet mount (100× magnification, oil immersion, phase contrast) (both A and B)**

FILAMENTS WITH SULFUR GRANULES

Using phase contrast and the 100× oil-immersion lens, sulfur granules can be observed within the cell. This often occurs in systems where reduced sulfur compounds such as hydrogen sulfide (H_2S) are present. Reduced sulfur compounds are found in septic wastewater and can also be found in some industrial waste such as paper (sulfite) mills.

Four filament types—Type 021N, *Thiothrix* spp., Type 0914, and *Beggiatoa*—are commonly seen with intracellular sulfur granules when reduced sulfur compounds are present. Although these filaments do appear often without sulfur granules, the presence of sulfur granules is an indication of waste septicity. These filaments are distinctly different from each other, and it is not critical that you positively identify each one. If you see any filament with sulfur granules, it doesn't matter which one because they all store sulfur for the same reason: reduced sulfur compounds or septic wastes are present.

Thiothrix Types I and II

The cells of *Thiothrix* Types I and II are barrel shaped. Sulfur granules are more readily seen in the smaller Type II. *Thiothrix* Type I is relatively large, and the sulfur granules appear much smaller (Figure 7-21).

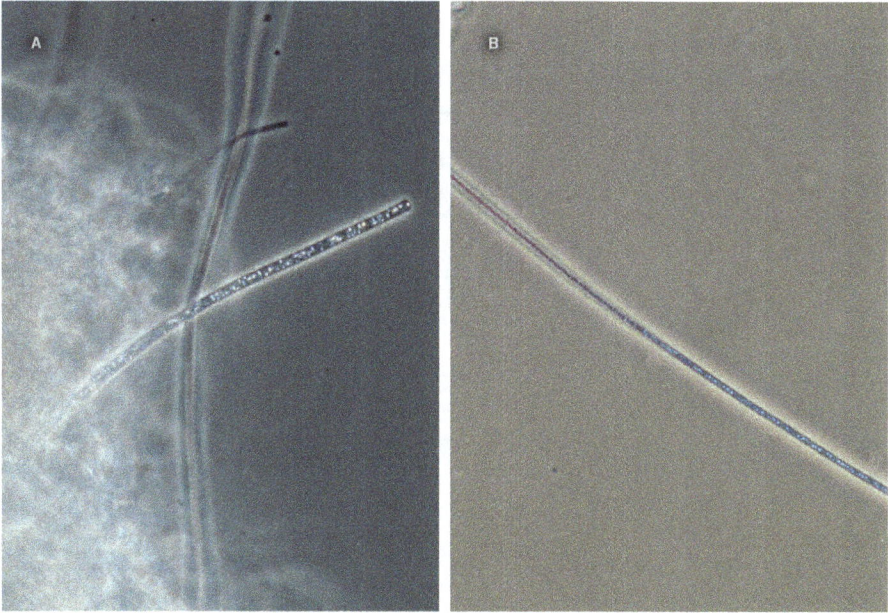

Figure 7-21 *Thiothrix* Type I (A) and *Thiothrix* Type II (B) with sulfur granules (wet mount, 100× magnification, oil immersion, phase contrast)

Type 0914

Type 0914 is similar in size to *Thiothrix* Type II but can be easily distinguished by its rectangular-shaped cells and rectangular-shaped sulfur granules that take up much of the filament space (Figure 7-22).

Type 021N

Type 021N is an unusually shaped filament. The cells are discoid shaped and look much like short, fat barrels or hockey pucks stacked on top of each other (Figure 7-23). Sulfur granules can be seen within the cells. The cells of type 021N stain gram-negative (Figure 7-24).

Beggiatoa

The individual cells of *Beggiatoa* are difficult to see. However, this filament is very easy to identify. Along with the presence of sulfur granules, *Beggiatoa* is motile; it glides slowly in the surrounding fluid (Figure 7-25).

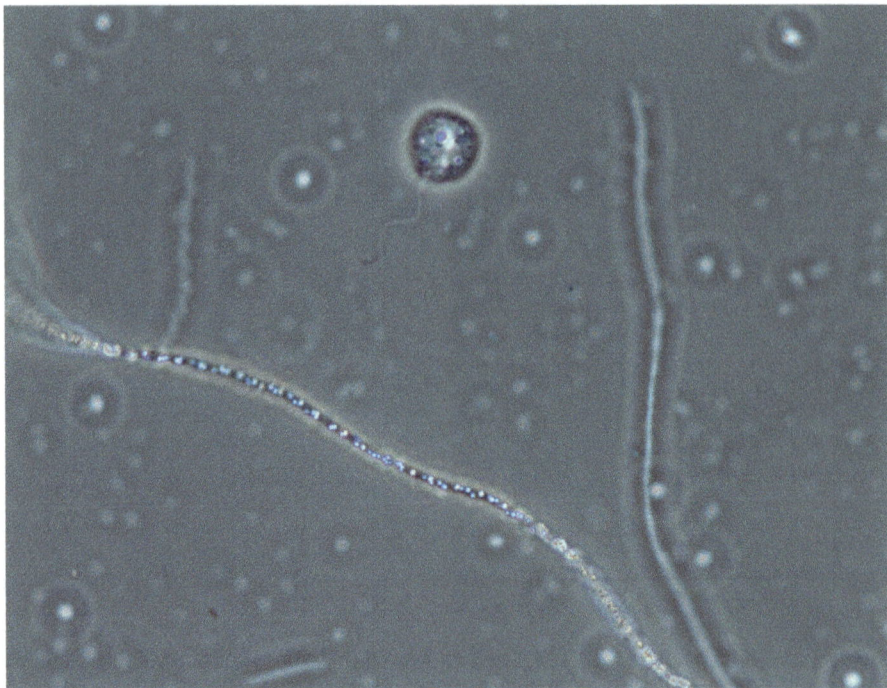

Figure 7-22 Type 0914 with sulfur granules (wet mount, 100× magnification, oil immersion, phase contrast)

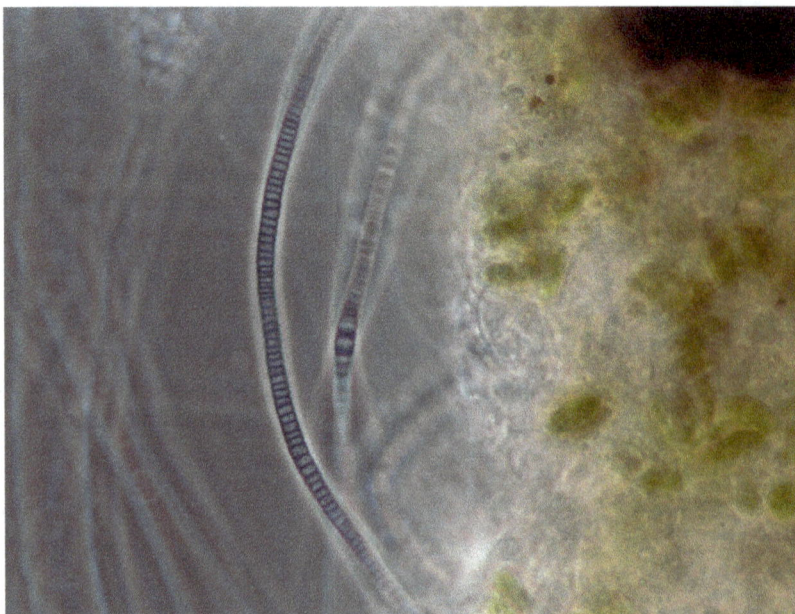

Figure 7-23 Type 021N with unique "hockey puck"–shaped cells (wet mount, 100× magnification, oil immersion, phase contrast)

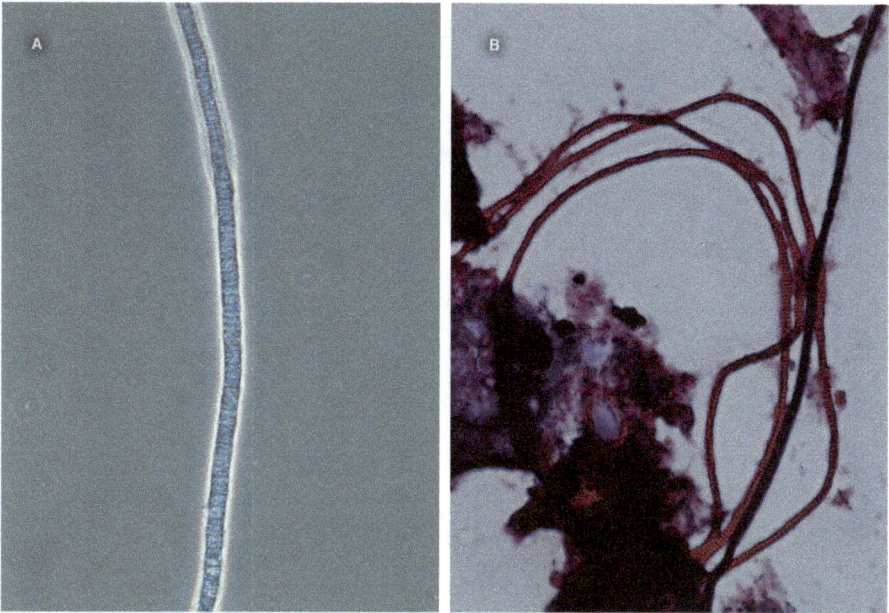

Source: Image A printed with permission from Glymph-Martin 2024

Figure 7-24 Type 021N with sulfur granules (100× magnification, oil immersion, phase contrast) (A) and gram-negative Type 021N (wet mount, 100× magnification, oil immersion, bright-field) (B)

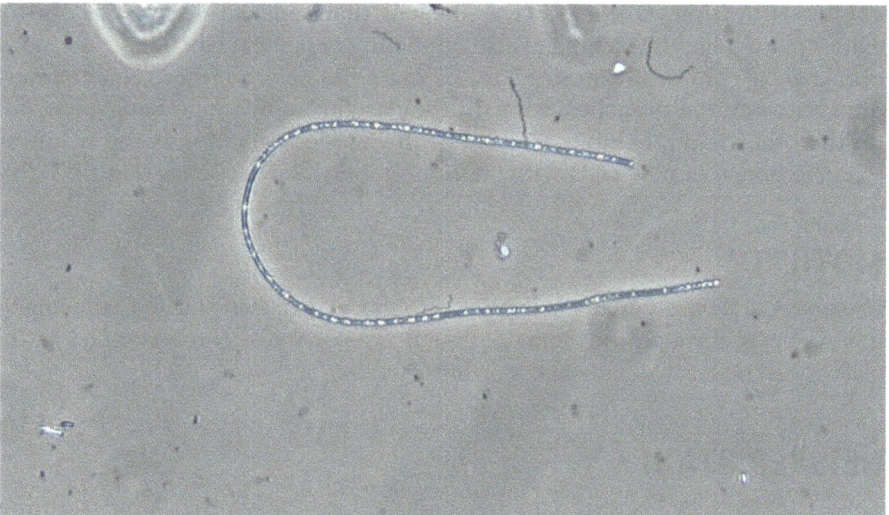

Figure 7-25 *Beggiatoa* with sulfur granules (wet mount; 100× magnification; oil immersion; phase contrast)

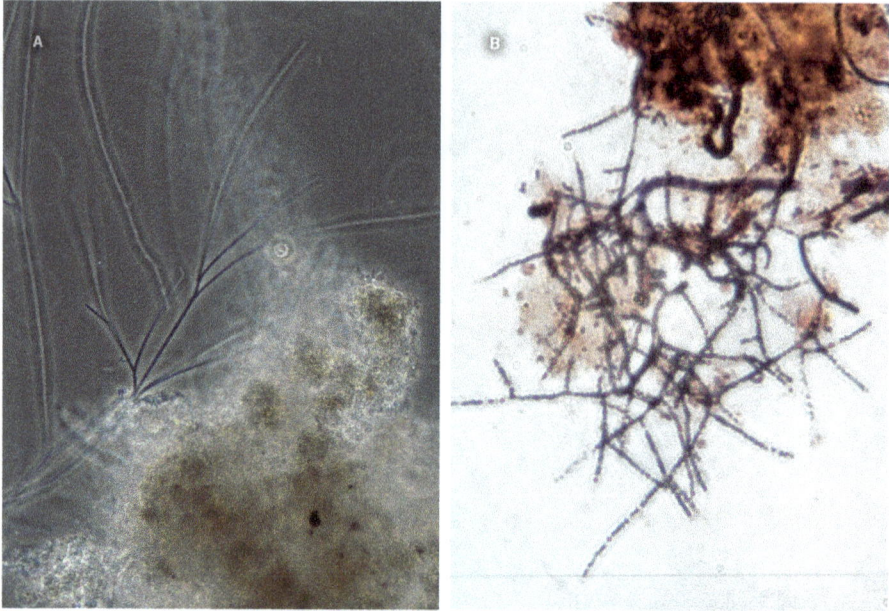

Figure 7-26 *S. natans* with false branching (100× magnification, oil immersion, phase contrast) (A) and *Nocardia* with true branching (Gram stain, 100× magnification, oil immersion, bright-field) (B)

BRANCHING FILAMENTS

There are two filamentous bacteria commonly seen in activated sludge that exhibit branching. *S. natans* is a filament that causes bulking and shows evidence of false branching. False branching appears when one or more branches are physically attached to the main filament but there is no continuum between the branches. It's as if they are just stuck together. *Nocardia*, on the other hand, exhibits true branching. True branching is the result of the emergence of new cells with continuum between branches (Figure 7-26).

MOTILITY

Beggiatoa is the only filament normally found in activated sludge that actually glides slowly through the fluid. Although it also stores sulfur granules in its cells, it can be easily spotted as it swims.

NEISSER STAIN

Most filamentous bacteria can be identified by observing Gram staining and other characteristics such as a sheath, attached growth, branching, sulfur granules, or branching. However, there are a few filaments that have a positive Neisser stain reaction. The Neisser stain identifies those filaments that have the ability to store phosphorus in the cell. This stain is most often used to identify phosphorus-accumulating organisms in biological phosphorus removal systems. This can be used as an additional tool to help identify filamentous bacteria. Five of the filaments that we have discussed have a positive Neisser stain reaction. Filament Type 021N, Type 0092, and *Nostocoida* are filaments that stain Neisser-positive (Figure 7-27 and Figure 7-28).

TYPE 0961

There is one filamentous bacteria type that is common to activated sludge that does not display any of the characteristics we have discussed thus far. It is not gram-positive or Neisser-positive. It does not branch or have a sheath or attached growth, nor does it store sulfur granules. However, it is very distinct and easy to recognize (Figure 7-29). This filament can thrive in conditions where there is very low soluble BOD present.

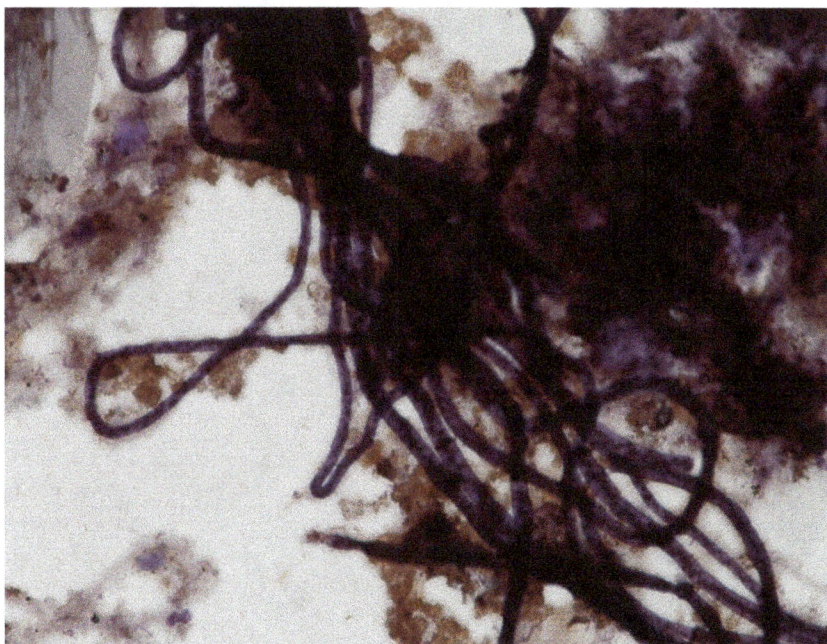

Figure 7-27 Type 021N, Neisser-positive (100× magnification; oil immersion; bright-field)

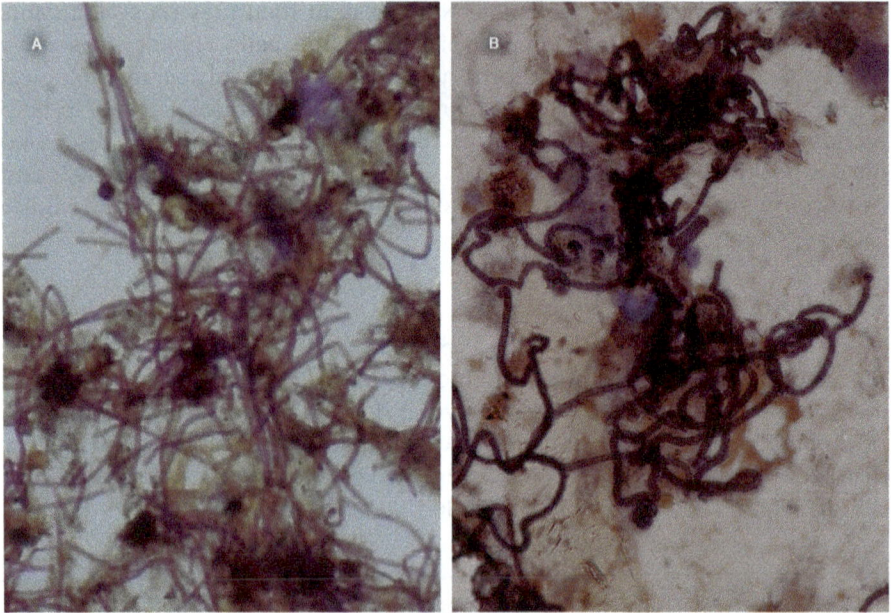

Source: Images A and B printed with permission from Glymph-Martin 2024

Figure 7-28 Neisser-positive Type 0092 (A) and *Nostocoida limicola* (B) (100× magnification, oil immersion; bright-field)

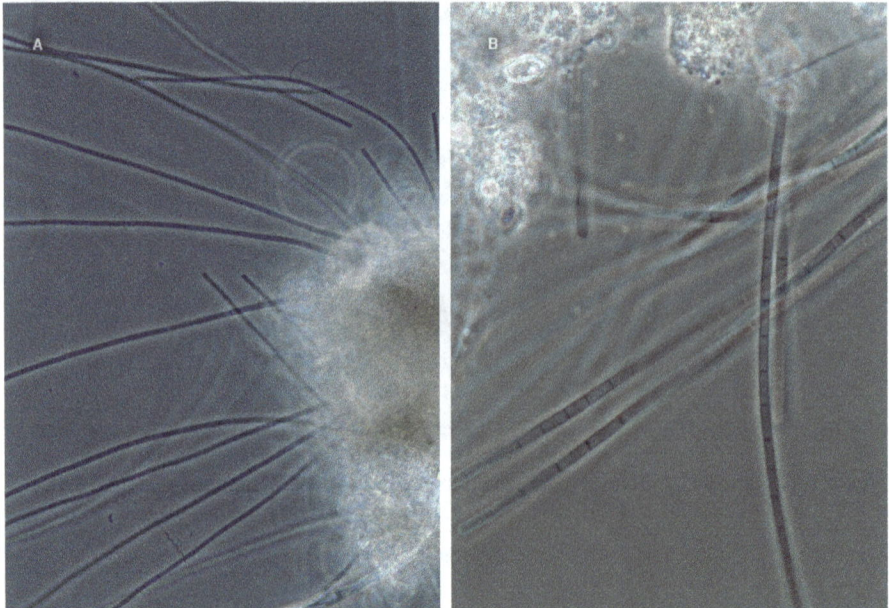

Figure 7-29 Type 0961 (100 magnification, oil immersion, phase contrast) (A and B)

Table 7-5   Gram-negative filamentous bacteria that cause sludge bulking

Filament Name	Appearance	Conditions That Can Favor Their Growth
Type 1701	Sheath: Sausage-shaped cells with long attached growth.	Low F/M ratio; low DO
S. natans	Sheath: Similar to Type 1701 but larger, with sausage-shaped cells. Exhibits false branching.	Low F/M ratio; low DO for the applied loading
H. hydrossis	Sheath: Very thin, straight filament. Protrudes from floc like pins in a pin cushion.	Low F/M ratio; low DO
Thiothrix Types I and II	Sheath and sulfur granules: Type I is a large, thick filament, and Type II is much thinner. Both have barrel-shaped cells.	Low F/M ratio; septic wastes; organic acids; wastes deficient in nitrogen
Type 0914	Sulfur granules: Thin, sheathed, square-shaped cells. Most often seen with square-shaped sulfur granules.	Low F/M ratio; septic wastes; organic acids; wastes deficient in nitrogen
Type 021N	Sulfur granules: Large filament with unique hockey puck–shaped cells.	Low F/M ratio; septic wastes; organic acids; wastes deficient in nitrogen
Beggiatoa	Sulfur granules and motile: Individual cell shapes are not visible but easy to identify, as Beggiatoa is the only one that swims.	Low F/M ratio; septic wastes; organic acids; wastes deficient in nitrogen; organic overloading

FILAMENT GROUPS

Some filamentous bacteria can be grouped together. These groups look alike and are often seen thriving under the same conditions. *S. natans* and Type 1701 form one such couple. Both have sausage-shaped cells in a tight-fitting sheath and generally thrive in low-DO conditions. The difference is that *S. natans* is larger and often exhibits false branching. Type 1701 is thin and often is seen with attached growth. However, if you encounter either and you are not sure which is which, it doesn't matter because they both are associated with low-DO conditions. Another filament group includes Type 0041, Type 0675, and Type 1851. All three thrive in low–F/M ratio and nutrient-deficient conditions, are gram-variable, have sheaths, have square-shaped cells, and are often seen with significant attached growth. The only difference is that Type 0041 is larger than Type 0675, and Type 1851 is the smallest. Should the widths of the filaments be measured to determine which it is? No. It really doesn't matter because all three thrive under the same conditions.

We can also group filaments together based on the conditions that favor their growth. For example, *S. natans*, *H. hydrossis*, and Type 1701 all thrive in low-DO conditions. Type 021N, *Thiothrix*, *Beggiatoa*, and Type 0914 all thrive in the presence of reduced sulfur compounds (septic wastes), organic

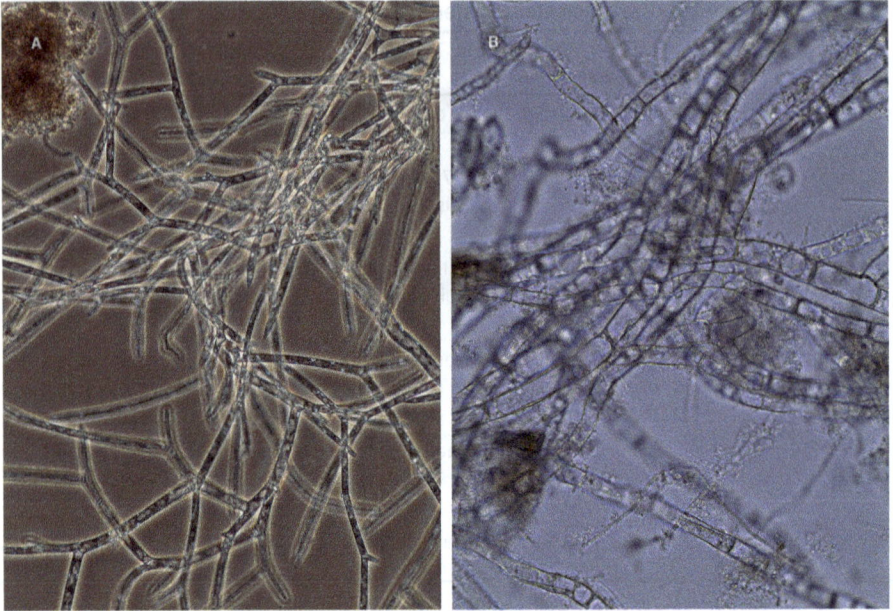

Figure 7-30 Fungi in activated sludge (20× magnification, phase contrast) (A); (40× magnification, phase contrast) (B)

acids, and waste deficient in nitrogen. Types 0041, 0675, and 1851 all thrive in low–F/M ratio and low-nutrient conditions.

FUNGI

Fungi are less common in activated sludge than filamentous bacteria. Although rare, Fungi can appear as large branching filaments that can dominate in the treatment system, causing severe bulking (Figure 7-30). Fungi have the ability to break down complex organic compounds in wastewater that are difficult for bacteria to degrade. Their dominance in the treatment system can affectively decrease the availability of nitrogen and phosphorus to bacteria, as they can easily assimilate these nutrients. Fungi are favored when the aeration basin pH level drops below 6.5 and in low–F/M ratio conditions.

CONTROLLING BULKING FILAMENTOUS BACTERIA AND FUNGI

Once filamentous microorganisms and fungi have been identified, determine the conditions that may be responsible for their dominance in the treatment system. A great deal of research has been conducted on the causes of filamentous bulking and foaming. The major conditions that have been associated with bulking and foaming are as follows:

- Low DO
- Low F/M ratio
- Longer sludge ages
- Septic wastewater; organic acids
- Nutrient-deficient incoming wastewater
- Changes in pH
- Excess FOG

Low DO

There are several filaments that are associated with low DO. If low DO is the issue, the operator can simply adjust the DO to meet the demand in the aeration basin. Sometimes, the DO level in the aeration basin is sufficient. In this case, look for areas of the plant where there is oxygen stress. These areas can be breeding grounds for filaments favored by low-DO conditions. For example, oxygen stress can occur in equalization tanks that are not aerated before discharge into the aeration basin or in aeration basins where there are dead zones caused by incomplete mixing. Maintaining high sludge blankets in clarifiers can also cause low DO in the system. However, for septic wastes that enter the treatment system, pre-aeration may be necessary. Septicity can result when incoming wastewater sits stagnant in pipes waiting to be pumped into the treatment system or can be introduced through septic wastes dischargers into the system. Increasing DO in the aeration basin during this time will help counter the impact from septic waste.

Low F/M Ratio

The F/M ratio is controlled through wasting. The influent loading (food) is not something the operator has much control over. However, through wasting, the operator has direct control of the microorganism (sludge) inventory. Increasing or decreasing wasting changes the relationship of the microorganisms to the food supply. The F/M ratio is directly proportional to wasting. When wasting increases, the F/M ratio will increase. The MCRT, sludge retention time (SRT), and sludge age are inversely proportional to wasting. In other words, the residence times increase as wasting decreases.

Long Sludge Age (MCRT and SRT)

Sludge age is the average time a particle of suspended solids remains in the treatment system. It refers to the amount of time it spends in the system. The calculation is based on the daily amount of solids entering the treatment plant and the total amount of solids in the aeration basin (mixed liquor suspended solids [MLSS]). Sludge age is based on the amount of solids in the aeration basin. Sludge age is calculated as follows:

$$\text{Sludge age (days)} = \frac{\text{Total pounds of MLSS in aeration basin}}{\text{Daily pounds of total suspended solids (TSS) in the influent}}$$

SRT is the average time cell mass stays in the treatment system. It is based on the suspended solids and on the amount of solids leaving the treatment system, including what is leaving the clarifiers. The SRT is calculated as follows:

$$\text{SRT (days)} = \frac{\text{Pounds/day of suspended solids in aeration basin}}{\text{Pounds/day of suspended solids wasted from the system}}$$

MCRT uses both the MLSS and the suspended solids to determine the time in days. It is a combination of what is in the system and what exits the system in the effluent as wasted suspended solids.

$$\text{MCRT (days)} = \frac{\text{Total pounds of MLSS in the secondary system}}{\text{Pounds/day of TSS wasted + pounds/day of TSS in the effluent}}$$

All three of these calculations represent how old the sludge is or how long it has been in the system. However, using the sludge age calculation is preferable and the simplest. It is the best representation of the amount of time the microorganisms remain in the aeration basin. The sludge age can be shortened by increasing wasting.

Septic Wastewater and Organic Acids

Septic wastewater is water that contains a significant amount of low-sulfur compounds as a result of anaerobic bacterial activity. To explain, when organic material is present, bacteria will use up the free oxygen first. Once all of the free oxygen is gone, they will use up the combined oxygen found in nitrates (that is why we see nitrogen gas/denitrification). Once that oxygen is used up, bacteria will use oxygen combined with sulfates. This results in the production of H_2S (reduced sulfur compounds). In other words, septicity results when all of the free oxygen and nitrate oxygen are used up. Wastewater septicity can result when sewage sits stagnant in pipes or when there is not enough oxygen present for the applied organic load and can result when treatment system receives significant loads from septage haulers.

Organic acids are produced by bacteria under anaerobic conditions. Organic acids can enter the aeration basin through anaerobic recycle streams such as anaerobic digester supernatant. Significant amounts of organic acids can also be produced in anaerobic zones used in biological phosphorus removal processes.

Septicity can be controlled by increasing the DO and/or pre-aeration before the aeration basin. Organic acids entering the aeration basin can be reduced by adding anaerobic digester supernatant slowly and consistently.

Creating an anoxic zone between the anaerobic zone and the aeration basin will help minimize the amount of organic acids entering the aeration basin.

Nutrient Deficiency

Generally, domestic wastewater contains sufficient nutrients. Industrial wastewater, as well as municipal systems with significant industrial inputs, may be lacking in nitrogen or phosphorus. Nutrient-deficient wastewater can be supplemented with the deficient nutrient. Determine the BOD/nitrogen/phosphorus ratio at the point where the wastewater enters the aeration basin. This will give a better measurement of the amount of nutrients available to the microorganisms. The amount of nutrients can be determined by measuring total nitrogen and total phosphorus entering the aeration basin. To ensure a healthy population of microorganisms, 5–10 mg/L of nitrogen and 1 mg/L of phosphorus are required for every 100 mg/L of BOD entering the aeration basin.

pH

The wastewater pH is completely dependent on the composition of the influent wastewater. Influent wastewater can contain a wide variety of chemicals and compounds that may influence the pH of the wastewater. In addition, alkalinity is lost during nitrification; therefore, it is important to maintain adequate alkalinity in the aeration tank to provide pH stability. Influent pH can be increased by adding either a caustic solution or a buffer solution to increase alkalinity.

FOG

Common sources of excessive FOG include restaurants, bars and grills, grocery stores, food-processing facilities, and homes. Most of the FOG are not soluble and do not mix well with water. Instead, FOG floats to the surface of basins and clarifiers and can even clog pipes. Some filamentous bacteria, particularly those that cause foaming, also float to the surface and have the ability to break down FOG and use them as a food source. The control of FOG must be done at the source by properly disposing grease in solid waste containers.

Primary treatment (primary clarification) plays a critical role in removing FOG before wastewater entering the aeration basin. Grease and oils float to the top of the primary clarifier and are skimmed off and removed for further treatment. Unfortunately, not all treatment systems have primary clarification.

REFERENCES

Eikelboom, D.H. 1977. "Identification of Filamentous Organisms in Bulking Activated Sludge." *Progress in Water Technology*, 8(2): 151–161.

Glymph-Martin, T. 2024. *Activated Sludge Filamentous Bacteria*. Wastewater Microbiology Solutions.

Jenkins, D., M. Richard, and G. Daigger. 1993. *Manual on the Causes and Control of Activated Sludge Bulking and Foaming*. Chelsea, Mich.: Lewis Publishers.

Chapter 8

Algae, Water Fleas, and Other Aquatic Worms

ALGAE

Algae are photosynthetic organisms that contain chlorophyll and obtain their energy from the sun and their carbon from carbon dioxide. There are three main classes of algae: brown algae (diatoms), green algae, and blue-green algae (cyanobacteria). Green algae are generally beneficial in lagoon treatment systems (Figure 8-1) because they supply oxygen for the bacteria (while bacteria supply carbon dioxide for algae) in the treatment process. Algae generally do not cause problems in activated sludge treatment systems. However, algae can be found growing in effluent channels, wherever there is stagnant water and on clarifier weirs. Proper cleaning and maintenance will help control algae growth.

WATER FLEAS

Water fleas (i.e., *Daphnia*) feed mostly on algae (Figure 8-2). So, if you see large numbers of water fleas skimming on the surface of the clarifiers, this is an indication that a significant amount of algae is present. If there are enough algae present and conditions are favorable (generally in spring and summer), water fleas can reproduce very rapidly. Often, they appear red as they swim on the surface of the clarifiers. *Daphnia* turn red when they produce hemoglobin, and they produce hemoglobin because there is not enough oxygen present in the water in the clarifier. This is an indication that the clarifiers need cleaning! The good news is that water fleas are very sensitive to various contaminants, so if there are a lot of fleas present, it means there are very few toxins in your wastewater!

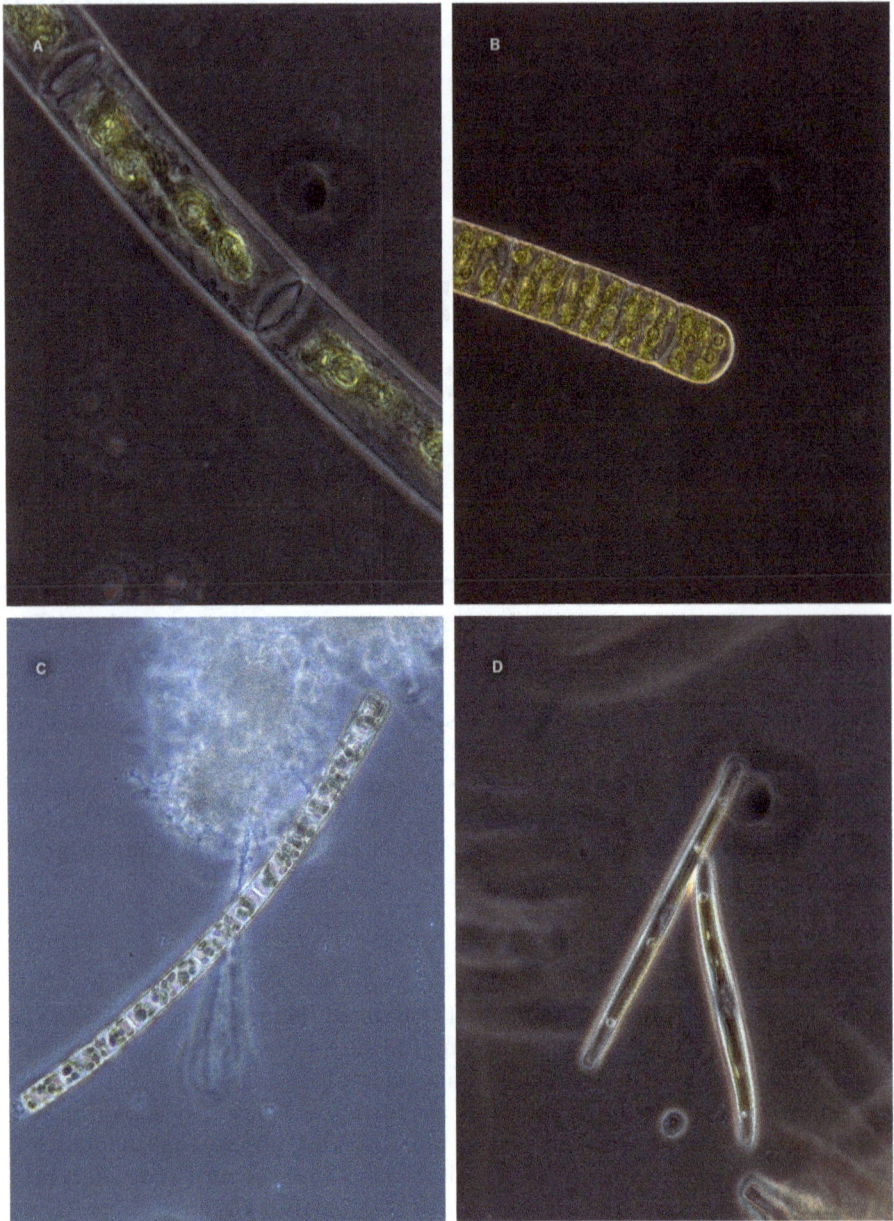

Figure 8-1   *Tribonema* (A) and *Oscillatoria* (B) (100× magnification, oil immersion; phase contrast); Ulothrix (C) and Synedra (D) (40× magnification, oil immersion, phase contrast)

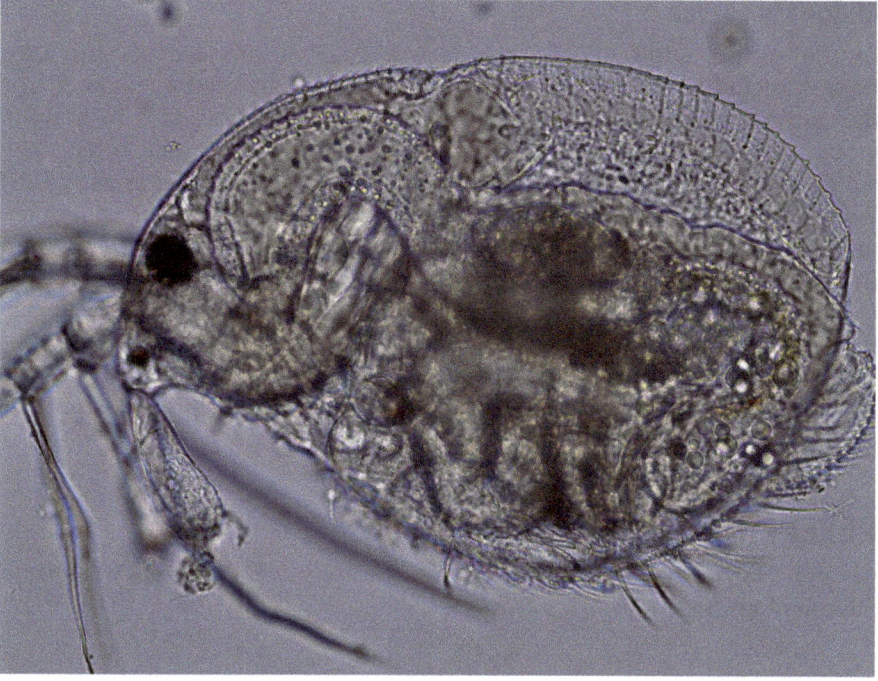

Source: Printed with permission from Glymph-Martin 2024

Figure 8-2 *Daphnia* **(10× magnification; phase contrast)**

OTHER AQUATIC WORMS

Tubifex Worms

Tubifex worms are often referred to as sludge worms or sewage worms. Where nematodes are microscopic and cannot be seen with the naked eye, tubifex worms are 10–20 times larger and can be easily seen with the naked eye. They are brown to pinkish in color and can turn pinkish to red in color when they produce hemoglobin in low-oxygen environments. The hemoglobin allows them to bind and store oxygen so that they can survive in hypoxic or anoxic conditions. They thrive in environments with high organic content and are often found in polluted or stagnant water, feeding on decaying organic matter, bacteria, and microscopic organisms. Tubifex worms burrow into sediments for protection, to access food, and for shelter from disturbances. These can grow in the activated sludge treatment system when sludge is held for a long period of time. The sludge satisfies their burrowing nature.

Midge Fly Larvae

In the spring, adult midge flies swarm over the surface of still water bodies, including secondary clarifiers. Midge flies lay their eggs on or near the surface of the water. The eggs hatch within a few days to a week. Once hatched, the larvae drop into the water in the clarifier, often using solids in the clarifier to form protective cocoons. They appear as reddish worms visible to the naked eye. They feed on organic matter, algae, and microorganisms in the sludge or water. They can also produce hemoglobin that enables them to survive in oxygen-poor environments. A quick fix is to install bug zappers in the spring to prevent them from swarming over the clarifier in the first place. Otherwise, there are products that will kill the larvae before they emerge. Make sure the product contains a bacterium called *Bacillus thuringiensis*. This particular bacterium produces specific toxins that target the larvae's digestive system.

REFERENCE

Glymph-Martin, T. 2024. *Activated Sludge Microbes Poster 2*. Wastewater Microbiology Solutions.

Chapter 9

Biological Phosphorus Removal

Phosphorus (P) is a major nutrient that is necessary for all living cells, including bacteria and other microorganisms. It is an essential element for living cells because of the critical role that it plays in biological processes. It is a key component in the structure of DNA and RNA molecules and in forming the backbone that supports our genetic code. Phosphorus is a necessary component of cell membranes and helps to maintain the pH balance within cells and tissues. It is also essential for the storage and transfer of energy within cells for various processes such as muscle contraction, cellular movement, and the synthesis of new cells. When bacterial cells need energy, one high-energy phosphate (PO_4) bond is broken to release energy. Overall, phosphorus is indispensable for life as we know it.

Phosphorus is also a nutrient required in the growth of aquatic plants, so the concentration of phosphorus in the waterways will determine the quantity of vegetative growth. When phosphorus is introduced into the receiving water, it can have an undesirable effect on the quality of the water.

Excess phosphorus can cause an overgrowth of aquatic vegetation and algae. There are several environmental problems associated with the rapid growth of aquatic plants. These include oxygen depletion, color, odor, taste, and turbidity issues, especially if the receiving water is used as source water for drinking. Consequently, environmental regulations are becoming more and more stringent in an attempt to minimize the discharge of phosphorus to the waterways.

Sources of phosphorus to the waterways include agriculture runoff. The phosphorus in animal manure and chemical fertilizers is necessary to grow crops. However, when nutrients such as phosphorus and nitrogen (N) are not fully used by plants, they can run off from the farm fields and negatively affect downstream water quality. Another source of phosphorus is stormwater. When precipitation falls on our cities and towns, it runs across hard surfaces such as rooftops, sidewalks, and roads, and it carries pollutants, including phosphorus, into the waterways. Yard fertilizers, pet waste, and certain soaps and detergents contain phosphorus and can contribute to nutrient pollution if not properly used or disposed of.

The average concentration of total phosphorus in municipal wastewater ranges from 10 to 20 mg/L. Although wastewater treatment systems are designed to treat large volumes of wastewater, these systems don't always remove enough nitrogen or phosphorus before discharging into waterways.

ABOUT PHOSPHORUS

Phosphorus exists in several forms. Elemental phosphorus in its pure form is very rare. It is usually found as a part of a phosphate molecule (PO_4). However, total phosphorus in aquatic systems exists in organic and inorganic forms. Organic phosphorus consists of a phosphate molecule associated with a carbon-based molecule (organic material). Phosphates that are not associated with organic material are inorganic. Both organic and inorganic phosphorus can either be dissolved in the water or suspended (attached to particles in the water column).

Inorganic forms of phosphorus include orthophosphates and polyphosphates. Orthophosphate makes up approximately 50–70% of the total phosphorus, whereas polyphosphates and organic phosphorus make up the remaining 30–50% (Gerardi 2006).

Phosphorus as orthophosphate is readily available for bacterial uptake, growth, and metabolism without any further breakdown. Polyphosphates, on the other hand, are complex molecules that break down very slowly, resulting in the release of orthophosphate. This breakdown (termed hydrolysis) can be chemically mediated or biologically mediated by bacteria and algae. Because of their stability in water, polyphosphates also easily sequester minerals such as aluminum, calcium, and iron.

Phosphorus tied to organic compounds is referred to as organic phosphorus. Organic phosphorous compounds are of minor concern in domestic wastewater.

When phosphorous compounds enter the activated sludge process, these compounds undergo biological and chemical changes. Some organic phosphorous compounds are removed from the wastewater when particulate, organic phosphorus is adsorbed to solids and settles out in the primary clarifier. In the aeration basin, with sufficient hydraulic retention time (HRT), organic phosphorus is degraded through microbial activity, resulting in the release of some orthophosphate in the aeration zone. Inorganic phosphorus in the form of polyphosphates is biologically and chemically hydrolyzed (broken down), and orthophosphate is also released in the aeration zone. When phosphorus is readily available (as organic P or *ortho*-P), phosphorus is removed from wastewater and incorporated or assimilated into cellular material as bacteria degrade substrate (soluble carbonaceous biochemical oxygen demand [cBOD]) and reproduce. The assimilated phosphorus makes up 1–3% of the bacterial weight (measured as mixed liquor volatile suspended solids [MLVSS]) (Figure 9-1).

Therefore, effluent phosphorus from the activated sludge process is approximately 90% orthophosphate. The orthophosphate may be present

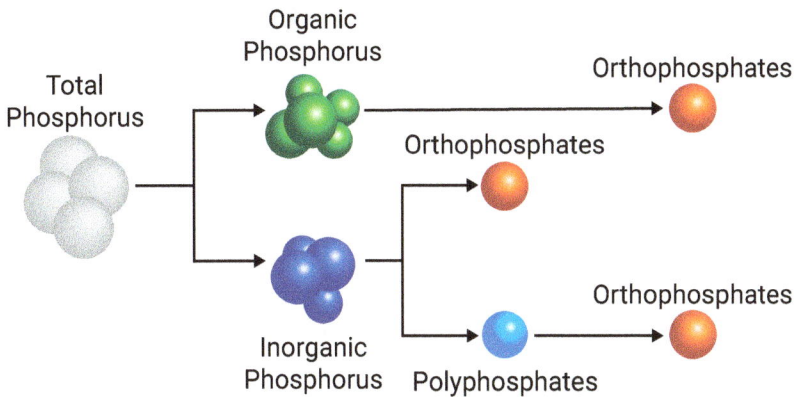

Figure 9-1 **Phosphorus compounds: Organic phosphorus and orthophosphate are both readily available for microbial uptake, whereas polyphosphate must be further hydrolyzed to orthophosphate before microbial uptake.**

as soluble ions or attached to solids. To reduce the concentration of effluent phosphorus from an activated sludge process, an advanced wastewater treatment measure is required. Phosphorus can be removed in municipal wastewater treatment plants through biological and chemical treatment measures. These measures include chemical precipitation of phosphorus and enhanced biological phosphorus removal (EBPR).

EBPR

EBPR is accomplished by certain groups of bacteria that occur naturally in the activated sludge treatment process. They enter the treatment process through feces in the wastewater and soil and from inflow and infiltration. These microorganisms, known as phosphorus (polyphosphate)–accumulating microorganisms (PAOs), are a group of bacteria that, under certain conditions, facilitate the removal of large amounts of phosphorus from wastewater. Biological phosphorus removal or "luxury uptake of phosphorus" occurs when the uptake of phosphorus by these bacteria is in excess of their normal cellular requirements.

The key to EBPR is the exposure of PAOs to alternating anaerobic and aerobic conditions. Therefore, a minimum of two treatment zones, an anaerobic (fermentative) zone and an aerobic zone, are required. The process involves hydrolysis and fermentation in the anaerobic zone and phosphorus uptake in the aerobic zone and incorporates the use of three groups of bacteria:

- Hydrolytic bacteria
- Fermentative bacteria
- PAOs

MICROBIAL ACTIVITY IN THE ANAEROBIC ZONE

Hydrolytic Bacteria

In the EBPR process, hydrolysis and fermentation occur in the anaerobic zone. Hydrolysis is basically the breakdown of complex substances to their simpler form or "building blocks." For example, proteins are hydrolyzed to amino acids, carbohydrates to simple sugars, and lipids to long-chain fatty acids. Hydrolytic bacteria have the ability to secrete extracellular enzymes called hydrolases (proteinase to break down protein, lipase to break down lipids, and glucosidases to break down carbohydrates). These simpler building blocks are soluble and can be readily absorbed into the bacteria cell for fermentation. So, the role of hydrolytic bacteria is to break down complex organic substances into simple soluble compounds for the fermentative bacteria to use (Figure 9-2).

Fermentative Bacteria

Fermentation is the microbial degradation of soluble organic compounds (cBOD) without the use of free molecular oxygen (O_2), bound oxygen from nitrate (NO_3^-) or sulfate, and carbon dioxide. Fermentation can also occur in anaerobic digesters, sewer systems, secondary clarifiers, and sludge thickeners.

There are two important groups of fermentative bacteria: the *acidogenic bacteria* and the *acetogenic bacteria*. The acidogenic bacteria, or acid formers, convert simple sugars, amino acids, and fatty acids to volatile fatty acids (VFAs; i.e., acetate, butyrate, lactate, and succinate) and other substances such as alcohols (i.e., ethanol and methanol), acetone, carbon dioxide, hydrogen, and water. Given enough time under anaerobic conditions

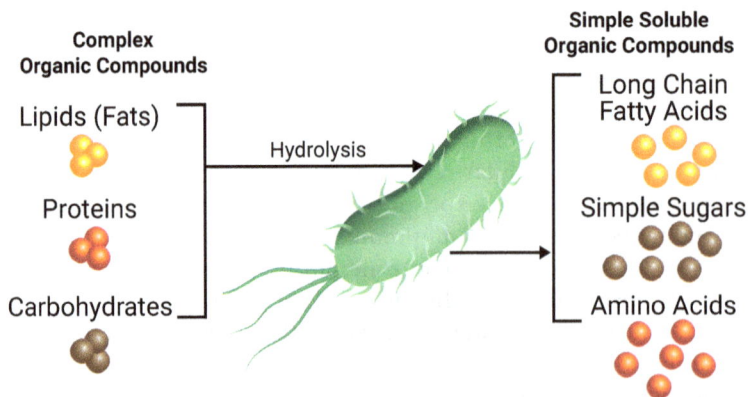

Figure 9-2 Hydrolysis: In the anaerobic zone, hydrolytic bacteria break down complex organic compounds into their simple soluble components by releasing enzymes; these soluble organic compounds are easily absorbed by the fermentative bacteria.

such as in an anaerobic digester, the acetogenic bacteria will convert the VFAs and alcohols into acetic acid, hydrogen, and carbon dioxide. For the purpose of the EBPR process, it is the availability of VFAs that is most important. This is one of the reasons why a proper detention time in the anerobic zone is important. I will discuss more about that later.

Fermentative bacteria can be both anaerobes and facultative anaerobes. Anaerobes are bacteria that do not live or grow when oxygen is present. Facultative anaerobes can live and grow with or without oxygen. Although facultative anaerobic bacteria and anaerobic bacteria are both capable of fermentation, the most important bacteria are the strict anaerobic bacteria. These bacteria also enter wastewater treatment plants through inflow and infiltration and fecal wastes.

During fermentation, VFAs are produced. These fermentative products are a necessary substrate for the PAOs. The presence of free oxygen or combined oxygen (nitrate) will interfere with the fermentative bacteria's ability to produce the VFAs that the PAOs require. Therefore, anaerobic conditions are a requirement for fermentation. It is important to understand that fermentation does not occur just because you have anaerobic conditions. Adequate carbon (organic compounds) must be present to support anaerobic bacterial growth and metabolism. So, the role of fermentative bacteria is to produce the VFAs that the PAOs require (Figure 9-3).

PAOs

Although PAOs are present in the anaerobic zone, these bacteria are strict aerobes, meaning that they require oxygen to grow. They cannot use (degrade) the VFAs that they require for phosphorus uptake under anaerobic conditions. The degradation of these compounds by PAOs occurs only in the presence of free molecular oxygen (in the aerobic zone) or nitrate (NO_3^-). Instead, in the anaerobic zone, PAOs rapidly absorb the VFAs, but they

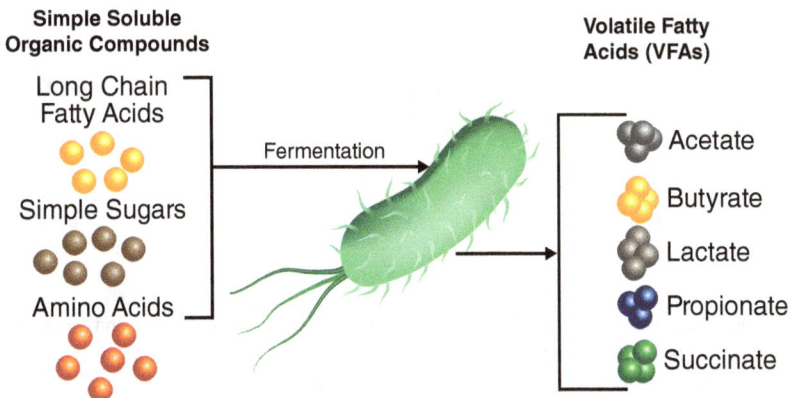

Figure 9-3 **Fermentation: Fermentative bacteria consume soluble organic compounds in the absence of oxygen and produce volatile fatty acids (VFAs).**

can only store them. They store them as insoluble starches called poly-β-hydroxyalkonates (PHAs). There are several different PHAs, but the most dominant one and the one that yields the highest phosphorus uptake is polyhydroxybutyrate (PHB). Henceforth, PHB will be used when referring to PHAs.

The process of storing PHB requires cellular energy. The energy needed is obtained through the breakdown of high-energy phosphate bonds already naturally present in the PAOs. Orthophosphate is then released to the bulk solution as PAOs gain the energy needed to absorb and polymerize the VFAs into PHB (Figure 9-4). So, *ortho*-P in the anaerobic zone comes from two sources: phosphorus already present in the influent wastewater and phosphorus that is released from the PAOs in the anaerobic zone.

PHB stored in the PAO cell in the anaerobic zone serves two important functions. Along with polyphosphates, PHB helps the aerobic PAOs to survive while in anaerobic conditions. Second, it helps PAOs to grow and replenish cellular polyphosphates by taking up soluble phosphate in the aerobic zone (when oxygen is present). Phosphorus uptake by PAOs in the aerobic zone is directly related to the amount of PHB that is stored in the cell

PAOs in the Anaerobic Zone

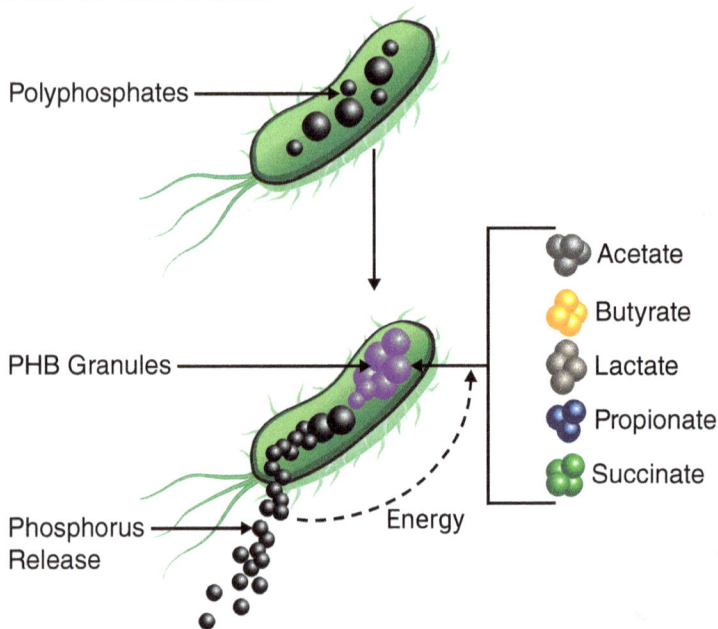

In the anaerobic zone, PAOs rapidly absorb the volatile fatty acids (VFAs) and store them as polyhydroxybutyrate (PHB) granules in the cell. The storage of PHB requires cellular energy. High-energy phosphate bonds are broken to provide the needed energy, and phosphorus is released into solution.

Figure 9-4   Phosphorus-accumulating organisms (PAOs) in the anaerobic zone

in the anaerobic zone. Therefore, it is critical to ensure that adequate storage of PHB occurs in the anaerobic zone.

However, there are other microorganisms that have the ability to store PHB. For example, PHB is a cellular component in floc-forming bacteria that aids in the formation of floc. Most critically, however, there are microorganisms in the anaerobic zone that also absorb VFAs and store them in the cell as PHB. The ones that are of most concern for the EBPR process are the glycogen-accumulating organisms (GAOs).

Glycogen-Accumulating Bacteria

In the anaerobic zone, the process of storing PHB requires cellular energy. PAOs gain energy through the breakdown of high-energy phosphate bonds and release of orthophosphate. GAOs, on the other hand, use their stored intracellular glycogen as an energy source for the storage of PHB (Figure 9-5). In the aerobic zone, where PAOs use energy from the stored PHB to take up phosphorus in excess amounts, GAOs do not take up phosphorus but, instead, use the stored PHB for growth and maintenance and to replenish

GAOs in the Anaerobic Zone

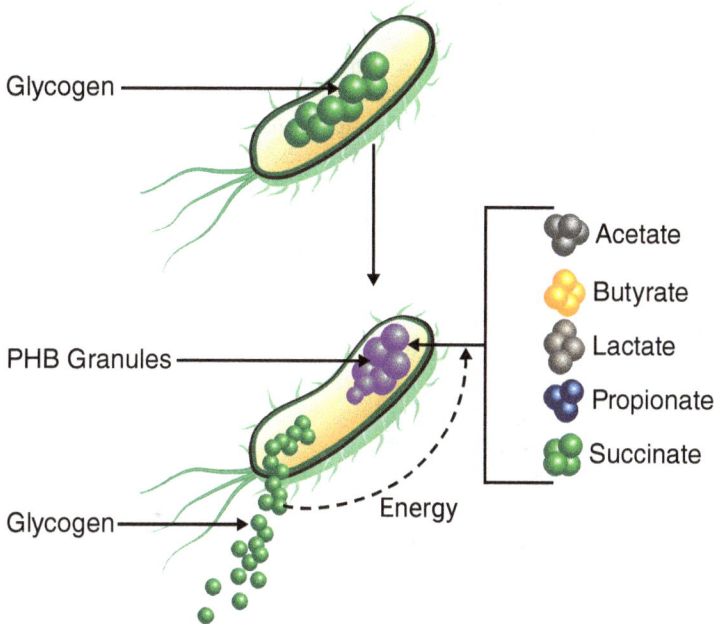

In the anaerobic zone, GAOs also rapidly absorb the volatile fatty acids (VFAs) and store them as polyhydroxybutyrate (PHB) granules in the cell. The storage of PHB requires cellular energy. Intracellular glycogen is used as an energy source for the storage of PHB. In this reaction, there is no phosphorus released into the bulk solution.

Figure 9-5   Glycogen-accumulating organisms (GAOs) in the anaerobic zone

glycogen. The presence of GAOs in the EBPR process is considered undesirable because they compete with PAOs for VFAs without contributing to the biological phosphorus removal process.

MICROBIAL ACTIVITY IN THE AEROBIC ZONE

The microbial activity in the aerobic zone of EBPR systems involves three groups of bacteria:
- Hydrolytic bacteria
- Floc-forming bacteria
- PAOs

Hydrolytic Bacteria in the Aerobic Zone

Hydrolytic bacteria are also present in the aerobic zone and serve the same purpose as those in the anaerobic zone. These bacteria hydrolyze complex organic compounds into their simpler forms so that they can be easily absorbed by the other bacteria in the aerobic zone, including floc-forming bacteria. In systems that have treatment components before the aeration basin, such as primary treatment or an anaerobic or anoxic zone, most of the complex compounds have been hydrolyzed before the wastewater enters the aerobic zone.

In EBPR systems, the bacteria population in the aerobic zone is made up of PAOs and floc-forming bacteria along with other heterotrophic bacteria that feed on organic substrate (carbon) entering the basin. However, remember that floc formers need carbon too. Too often in EBPR systems, the focus is on making sure enough carbon is available in the anaerobic zone to facilitate VFA production. However, residual carbon is needed in the aeration basin for the floc-forming bacteria. In a well-operated EBPR system, a significant portion of the bacteria in the aerobic zone are PAOs. Therefore, the carbon requirement in the aerobic zone of EBPR systems is lower than that in traditional activated sludge systems. In the aerobic zone, in the presence of oxygen, PAOs use stored PHB as a food source. Very little additional carbon is required for them in this zone. So, in essence, PAOs enter the aerobic zone with their own stored food source. Therefore, most of the carbon entering the aerobic zone will be used by the floc formers.

It is important to note that although PAOs can grow in clusters, they are not floc-formers, so they need floc particles to attach to. Therefore, it is also important to maintain a healthy population of floc-forming bacteria. Also, PAO cells are denser and settle faster, so the sludge volume index in EBPR systems with a healthy population of PAOs is generally lower than that in traditional activated sludge.

PAOs in the Aerobic Zone

In the aerobic zone, the presence of free oxygen enables the PAOs to degrade the stored PHB as a carbon and energy source. At the same time, the PAOs absorb *ortho*-P using the energy they gained while degrading PHB. The PAOs obtain so much energy from the degraded PHB that they absorb not only enough to replace the released phosphorus but also additional large quantities of the influent phosphorus (Figure 9-6).

The absorbed phosphorus is stored as polyphosphate granules or volutin. Phosphorus, which is stored in bacteria cells as intracellular volutin granules or polyphosphates, may be 1–3% of the dry weight of a bacterium. In addition, the uptake of phosphorus renews the polyphosphate pool in the bacteria in the return sludge so that the process can be repeated again. Phosphorus removal from the system is achieved when the bacteria (sludge) are wasted from the secondary clarifier. Sludge that is not wasted is returned to the anaerobic zone, where the EBPR process is repeated.

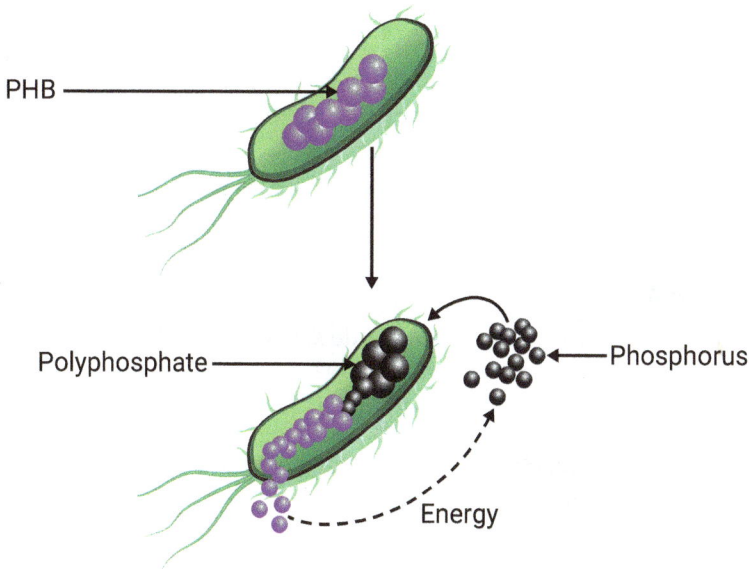

In the aerobic zone, where free oxygen is present, PAOs quickly consume the stored polyhydroxybutyrate (PHB). The energy gained from the breakdown of the PHB is used to absorb phosphorus in excess of cellular need. The phosphorus is stored in the cell as polyphosphate granules.

Figure 9-6   Phosphate-accumulating organisms (PAOs) in the aerobic zone

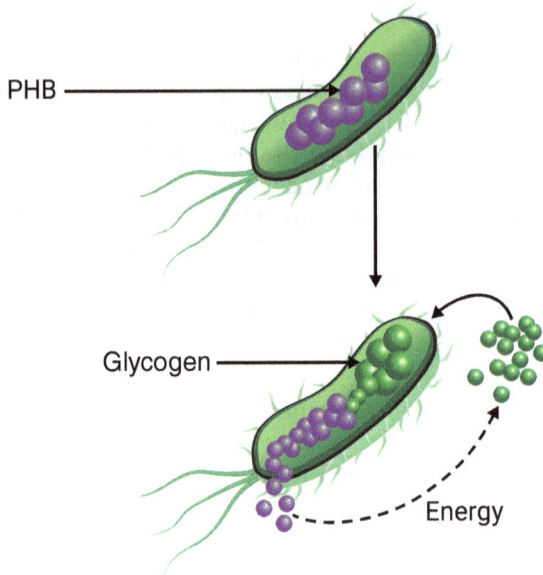

In the aerobic zone, where free oxygen is present, GAOs quickly consume the stored polyhydroxybutyrate (PHB). The energy gained from the breakdown of the PHB is used for cell maintenance and growth and to replenish the glycogen used up in the anaerobic zone. GAOs in the aerobic zone do not contribute to phosphorus uptake.

Figure 9-7   Glycogen-accumulating organisms (GAOs) in the aerobic zone

GAOs in the Aerobic Zone

GAOs are a mixed culture that can include several bacterial strains. However, they all have the same metabolism. GAOs have the capacity to use stored glycogen to take up soluble carbon substrates (VFAs) anaerobically and store them as PHB. In the aerobic zone, GAOs use the stored intracellular PHB to produce more intracellular glycogen. In the aerobic zone, GAOs do not contribute to phosphorus removal. Instead, they metabolize PHB to gain energy for glycogen formation (Figure 9-7).

FACTORS THAT AFFECT EBPR PERFORMANCE

Carbon

All organic material contains carbon. So, if carbon is required, it simply means that organic compounds are required. Carbon (organic) compounds are biodegradable, which means microbes can use these compounds for growth and cell maintenance. However, bacteria can only use (absorb) soluble (dissolved) organic compounds. Some carbon compounds entering the treatment system are readily biodegradable, meaning that they are made up of simple soluble molecules and can be easily absorbed into the bacteria cell. Others, although still biodegradable, are made up of more complex (particulate and colloidal) compounds that cannot be readily absorbed. These complex molecules must first be hydrolyzed to simple soluble compounds before they can be absorbed into the bacteria cell.

As discussed earlier, microbial activity in the anaerobic zone of the EBPR system takes place in two stages. The first stage is called "hydrolysis." Hydrolysis involves the breakdown of complex substances into simplistic, soluble organic compounds that can be absorbed by bacterial cells. The second stage is "fermentation." Once the soluble compounds are used, degradation results in the production of VFAs and alcohols (Figures 9-2 and 9-3).

How does Raw Wastewater Carbon Composition Affect VFA Production?

Raw wastewater that is composed mostly of soluble organic compounds has readily available substrate that can be immediately used by fermentative bacteria to quickly produce the VFAs. If the wastewater is composed of mostly complex organic compounds, these have to be hydrolyzed first to soluble organic compounds before they can be used by the bacteria. Therefore, more time is required to produce the needed VFAs. So, the composition of the wastewater matters. The average composition of raw wastewater is shown in Table 9-1.

Keeping the data in mind, only 75.4% of the substances entering wastewater treatment systems are available for microbial uptake. Of that amount, only 17.5% is readily biodegradable and can be used immediately for the production of VFAs. The other 57.9% has to be hydrolyzed before it can be absorbed by the microorganisms.

Wastewater treatment operators have limited control over the composition of the wastewater entering the treatment plant. However, operators do have some control over the amount of time allowed for the hydrolysis of complex compounds by adjusting the retention time in the anaerobic zone. The more complex the wastewater composition is, the more time will be required for hydrolysis, fermentation, and the production of the necessary VFAs.

Table 9-1   Typical wastewater composition

Type of Compound	Description
Inert particulates (17.1%)	These are particulates that are not biodegradable and cannot be used by the microorganisms.
Complex organic compounds (57.9%)	These are complex compounds that are slowly biodegradable and must first be hydrolyzed to simple, soluble organic compounds before they can be absorbed by the microorganisms.
Inert soluble compounds (7.8%)	Although these are soluble compounds, they are not biodegradable and cannot be used by the microorganisms.
Simple organic compounds (17.5%)	These are compounds that are readily biodegradable and that can be immediately absorbed and used by the microorganisms.

Source: Pasztor et al.2009

VFAs

VFAs are the simplest form that organic material can be broken down into. Generally, VFAs are only present in small amounts in the influent wastewater; thus, additional VFAs are needed for the system to work properly. Therefore, the role of the anaerobic zone is to simply provide additional VFAs. These VFAs are the "life blood" of the EBPR process. Without them, PAOs will not store PHB in the anaerobic zone, and without PHB, PAOs will not take up phosphorus in excess in the aerobic zone.

In the EBPR system, complex organic compounds are hydrolyzed into simple compounds that can be absorbed by the bacteria. Fermentative bacteria use these simple compounds, and VFAs are produced. A mixture of different VFAs is produced in the anaerobic zone. They include butyrate, acetate, propionate, succinate, isobutyrate, valerate, and isovalerate.

The three main VFAs—produced as a result of degradation of protein, carbohydrates, fats, and other organic compounds in wastewater—are acetate, butyrate, and propionate. Acetate is found to be the least toxic of all VFAs, followed by butyrate and propionate. However, acetate can sometimes favor the growth of GAOs over PAOs, which may lead to the failure of EBPR. Therefore, even though propionate is generally more toxic than acetate, it is preferred over acetate.

Synthetic VFAs can be added but are expensive. So, as an economical solution, VFAs can be produced on-site through the anaerobic digestion of sludge and later introduced to the treatment steps in the EBPR process.

What Factors Affect VFA Production in the Anaerobic Zone?

The amount of VFAs generated depends on several factors: wastewater composition, pH, mixed liquor concentration, and anaerobic–aerobic HRT and solids retention time (SRT).

Wastewater composition. As previously mentioned, VFAs are the direct result of the fermentation of soluble carbon compounds. Wastewater that contains an appreciable amount of soluble, simple organic compounds has substrates that are readily available to microorganisms. Fermentation and the production of VFAs can proceed rapidly. On the other hand, fermentation and VFA production proceed much slower in wastewater that contains an appreciable amount of complex organic compounds. The more complex the wastewater, the more time will be required to produce adequate amounts of VFAs.

Mixed liquor solids concentration. In the anaerobic zone, hydrolysis occurs to break down complex organic compounds into soluble organic compounds. These soluble compounds are required for fermentation and the production of VFAs to occur. Hydrolysis is mediated by heterotrophic bacteria. Biological floc in the return activated sludge (RAS) is made up of heterotrophic bacteria, so the rate of hydrolysis is affected by the mixed liquor concentration in the anaerobic zone. Hydrolysis increases with increasing mixed liquor concentration. In addition, the mixed liquor in the aerobic zone, where phosphorus uptake occurs, contains PAOs with stored polyphosphate. These PAOs in the return sludge must come in contact with the VFAs in the anaerobic zone so that phosphorus release and the storage of PHB can occur.

It is important to note that more is not necessarily better. Increasing the mixed liquor concentration beyond a certain point can be problematic. If carbon is available and too many VFAs are produced, excess VFAs can promote the growth of GAOs. In addition, excess VFAs spilling over into the aeration basin can adversely affect the activated sludge process (see Chapter 11). Also, if the mixed liquor concentration is too high, there may not be enough carbon available in the influent to produce enough VFAs for the number of PAOs present in the return sludge. This can also cause die-off (because of a lack of food) and autolysis (the destruction of cells). During the process of autolysis, additional phosphorus is released. The average mixed liquor concentration in the anaerobic zone can range from 1,500 to 3,000 mg/L, depending on the EBPR process configuration.

pH. The pH in the anaerobic zone can be influenced by many different factors. Various chemicals compounds and even biological processes can cause changes in pH. Most importantly, in EBPR systems, pH can affect anaerobic P release and aerobic P uptake rates, as well as the P release/carbon ratio. The P release rate increases as pH increases up to about a pH of 8.0. Above a pH of 8.0, the P release rate will begin to decrease. In addition, pH values greater than 7.5 favor PAOs over their competitor GAOs.

SRT and HRT. The growth rate for PAOs is relatively fast. These bacteria multiply approximately every 20 min. Ideally, the SRT should be based on the influent carbon loading. If the SRT is too long and the carbon loading is too little, there will not be enough VFAs produced to sustain the PAO population in the anaerobic zone. This will reduce the PAO population and thereby reduce EBPR performance. In addition, if sufficient carbon is present, significant amounts of phosphorus will be released in the anaerobic zone. This will increase the phosphorus loading to the aerobic zone and reduce the capacity for the uptake of influent phosphorus in the aerobic zone.

Although the SRT can affect the EBPR process, the HRT is the most critical. In the anaerobic zone, time must be allowed for hydrolysis, the production of VFAs, and for PAOs to take up the available VFAs, release phosphorus, and store PHB. Keep in mind that PAOs are aerobic microorganisms. The stored PHB helps to maintain them under anaerobic conditions. If the HRT is too long, they will begin to degrade the stored PHB to survive anaerobic conditions. Without PHB, the PAOs cannot take up phosphorus when they enter the aerobic zone.

The longer the bacteria remain in the water in the anaerobic zone, the more carbon they will consume, and additional phosphorus (secondary P release) can be released.

Other Factors Affecting Performance

Other factors that affect the EBPR process include temperature, cations con-centration, and nutrients.

Temperature

PAOs are lower-range mesophiles and/or psychrophiles. Mesophiles grow best at moderate temperatures between 20 and 45°C. Psychrophiles grow in colder temperatures (<20°C). Because PAOs prefer cooler temperatures, they grow best when the wastewater temperature is between 20 and 25°C but can still grow well at temperatures that are even cooler. Their competitors, GAOs, are midrange mesophiles with an optimal temperature range from 25 to 35°C. Studies have shown that optimal PHB storage occurs around 35°C. This gives GAOs the advantage at warmer temperatures. They can easily outgrow PAOs during the summer months when VFAs are available.

Cations

Cations present in the wastewater assist in phosphorus uptake and the stor-age of intracellular orthophosphate. They play a role in the binding mecha-nisms of phosphorus and in the stabilization of intracellular polyphosphate. The concentration and composition of cations in the influent wastewater play an important role in maintaining the stability of the EBPR process.

Three cations are the most important: magnesium, potassium, and cal-cium. Just like phosphorus, magnesium and potassium are released in the anaerobic zone and taken up in the aerobic zone. Municipal wastewater typically contains an adequate supply of these cations. Occasionally, they may be deficient.

How much are required? For every 10 mg/L of P removed, bacteria require (Pattarkine and Randall 1999)
- 5.6 mg/L: magnesium;
- 6.3 mg/L: potassium; and
- 3.2 mg/L: calcium.

Nutrients

Nutrient deficiency is a common occurrence in activated sludge processes. Most of the time, it is caused by the presence of nutrient-deficient industrial wastewater. Nutrients that are most often deficient are nitrogen and phos-phorus. Soluble substrates that are absorbed by bacterial cells cannot be degraded properly without sufficient amounts of nutrients. Instead, much of the substrate is converted by bacterial cells to insoluble polysaccharides and stored outside the bacterial cells. These polysaccharides are viscous and slimy and can cause severe bulking and even foaming in the aerobic zone.

Nutrient deficiency can be prevented by ensuring that the necessary quantities of nutrients are present in the influent to both the anaerobic and the aerobic zones. Remember, for every 100 mg/L of carbon that is con-sumed by bacteria, they need 5–10 mg/L of nitrogen and 1 mg/L of P.

Secondary Phosphorus Release

When mixed liquor solids, rich in phosphorus, are allowed to become anaerobic in areas outside of the anaerobic zone, secondary phosphorus release can occur. This can occur in final clarifiers, RAS wet wells, pipelines, and digesters. Secondary phosphorus release can also occur in the anaerobic zone if the detention time is too long.

MONITORING EBPR SYSTEMS

Monitoring Specific to the Anaerobic Zone

While the EBPR process may seem complex, the steps to control the process are relatively straightforward. Management of the environment involves creating anaerobic and aerobic conditions in the proper sequence and managing return sludge and mixed liquor recycle flows. Managing the food source involves ensuring proper nutrient ratios, sufficient amounts of carbon, and sufficient VFAs to promote enhanced phosphorous uptake by the organisms. Routine monitoring will ensure that the parameters are within the operational values for EBPR performance.

If the system is operating properly, monitoring in the anaerobic zone should show the following:
- an increase in *ortho*-P from influent to anaerobic zone effluent
- a decrease in oxidation–reduction potential (ORP) values from influent to effluent
- ORP in the range of –100 to –300
- PHB granules observed using the PHB stain

Monitoring every parameter may not be practical; however, the more information you can collect, the better able you will be to monitor, correct, and troubleshoot the EBPR process. Table 9-2 contains parameters that should be monitored in the anaerobic zone.

Monitoring Carbon

The purpose of carbon in the EBPR system is to provide substrate in the anaerobic zone that can be used by the anaerobic bacteria to produce

Table 9-2 Process monitoring in the anaerobic zone

Chemical Parameters	Process Control Parameters
• Carbon (BOD and/or chemical oxygen demand [COD])	• HRT
• VFAs	• SRT
• Phosphorus	• Dissolved oxygen (DO)/ORP
• pH	• Solids mass fraction/food-to-microorganism (F/M) ratio

BOD—biological oxygen demand, COD—chemical oxygen demand, DO—dissolved oxygen, F/M ratio—food-to-microorganism ratio, HRT—hydraulic retention time, ORP—oxidation–reduction potential, SRT—solids retention time, VFA—volatile fatty acid volatile organic acids. The amount of volatile acids required depends on how much phosphorus must be removed. Unfortunately, the amount of carbon in raw sewage entering the plant is beyond the operator's control. Carbon supplements are available, but costs must be weighed. Keep in mind that more is not necessarily better. Carbon/P ratios greater than or equal to 50/1 favor the growth of GAOs.

How much carbon is required? Approximately 30 to 40 mg/L of BOD (> 45mg/L COD) is required to produce enough VFAs to remove 1 mg/L of P (BOD—biological oxygen demand, COD—chemical oxygen demand, TP—total P).

Supplemental substrate may be added to the anaerobic/fermentation zone to increase COD/BOD loading if necessary. Primary or return sludge can also be used to provide additional carbon. The total organic substrate in return active sludge can range from 4 to 30% dry weight.

Monitoring carbon using the COD test. BOD measures the amount of DO required by aerobic organisms to break down biodegradable organic material present in the wastewater. The test is used to measure the organic strength of the wastewater. The rate of oxygen consumption depends on the temperature, pH, microorganisms present, and type of organic material in water.

COD measures the amount of DO required for the chemical oxidation of both organic and inorganic chemicals like ammonia and nitrite. This method is most often used to measure the amount of industrial waste in wastewater, which cannot be measured using the BOD test.

The COD test uses a chemical (potassium dichromate in a 50% sulfuric acid solution) that "oxidizes" both organic (predominant) and inorganic substances in a wastewater sample, which results in a higher COD concentration than BOD concentration for the same wastewater sample; however, for EBPR systems, BOD is a more accurate measure of the biodegradable material that will be available to microorganisms in the anaerobic zone for the production of VFAs.

Although BOD, particularly cBOD, is a more accurate measure of the organic content in the wastewater, using this measure for process control in some cases may not be practical because the test takes five days. The COD test, however, can be completed in 2–4 hours and is acceptable to use for process control. However, the relationship between COD and BOD in wastewater must be monitored.

The relationship between BOD and COD. COD is normally higher than BOD because more organic compounds can be chemically oxidized than biologically oxidized. Normally, the COD is in the range of 1.3–1.5 times higher than the BOD value. When COD value is more than twice the BOD value, there is good reason to suspect that a significant portion of the material in the wastewater is chemical and not biodegradable by the microorganisms in the wastewater.

The ratio of BOD/COD in wastewater is a good indication of how much of the total organic load (or oxygen demand) is available for biological degradation by the microorganisms. The lower the ratio, the more nonbiodegradable matter it contains.

Typical values for the ratio of BOD to COD for untreated municipal wastewater are in the range of 0.3–0.8. A ratio of 0.5 or higher typically means the wastewater is more easily biodegradable by the microorganisms (easily biologically treated).

Also, whether you are measuring COD or BOD, PAOs are sensitive to disturbance and require stable organic loading. Significant swings in COD or BOD loading can adversely affect EBPR operations. It is always best to "equalize" the influent to the anaerobic/fermentation zone and to make every effort to prevent slug discharges. Any changes in loading rates should be implemented gradually over an extended period of time.

Where to monitor for BOD and/or COD.
- Anaerobic zone influent: to ensure sufficient carbon is present for the production of required amounts of VFAs

Monitoring VFAs

How much VFAs is required? The amount of PHB stored in the PAO cell in the anaerobic zone is directly related to how much phosphorus uptake will occur in the aerobic zone. Mostly, VFAs are generated in the anaerobic zone by microbial activity. Approximately 7–10 mg/L of VFAs is used to store enough PHB to remove 1 mg/L of P. However, if at least 7 mg/L of VFAs remains in the discharge of the anaerobic zone, it is assumed enough was available for the microbes (if you got some left over, there must have been enough).

Where to monitor for VFAs
- Anaerobic zone effluent: to ensure sufficient VFAs were present for the storage of required amounts of PHB

Monitoring Phosphorus

The amount of phosphorus in the influent wastewater determines how much carbon is required as well as how much VFAs are needed, and it influences other parameters such as the HRT/SRT and the mixed liquor concentration in the anaerobic zone. It is equally important to monitor phosphorus in the anaerobic zone effluent. It is important to monitor the amount of phosphorus entering and exiting the anaerobic zone.

When PAOs uptake VFAs, the energy required to convert and store them as PHB is gained from the breakdown of high-energy phosphate bonds. As a result, *ortho*-P is released. So, the amount of phosphorus leaving the anaerobic zone should be greater than the amount of phosphorus entering the zone. The amount of phosphorus released can range from 5 to 50 mg/L. The release of phosphorus in the anaerobic zone signals that VFA uptake and PHB storage have occurred.

The difference between the anaerobic zone influent and effluent phosphorus concentration is assumed to estimate the amount of phosphorus released. Although the release of phosphorus is critical to the EBPR process, if too much phosphorus is released in the anaerobic zone, it will add to the total amount of phosphorus that must be removed in the aerobic zone and may exceed the capacity of the PAOs that are available to remove it.

Monitoring for TP versus ortho-P. The TP test measures all forms of phosphorus in the sample. Most effluent limits in discharge permits are for TP for the purpose of determining the overall reduction of phosphorus in the treatment system. However, the phosphorus released by PAOs in the anaerobic zone is in the form of orthophosphate. Monitoring orthophosphates in the anaerobic zone is a more direct measure of the amount of phosphorus released in the zone.

Where to monitor for phosphorus.
- Anaerobic zone influent
  - TP: Monitoring TP will help determine the level of treatment required.
  - *Ortho*-P: Monitoring *ortho*-P will provide a baseline for the amount of *ortho*-P entering the anaerobic zone.
- Anaerobic zone effluent
  - *Ortho*-P: Comparing influent *ortho*-P with anaerobic zone effluent *ortho*-P will provide a measure for the amount of phosphorus released by the PAOs.

Monitoring pH

Because pH affects the VFA production, VFA uptake rate, and the rate of phosphorus release, it is important to monitor pH in the anaerobic zone.

Where to monitor for pH.
- Anaerobic zone: to monitor effects of pH on VFA production, VFA uptake, and phosphorus release

Monitoring Anaerobic Zone HRT

The anaerobic zone HRT is a measure of the average length of time that the wastewater remains in the anaerobic zone. The wastewater must stay in the zone for an adequate period of time to produce the necessary VFAs and to allow for the uptake and storage of PHB. The HRT is calculated as follows:

$$\text{Detention Time (hours)} = \frac{\text{Tank Volume } \left(\text{ft}^3\right) \times 7.48 \left(\text{gallons/ft}^3\right) \times 24 \text{ (hours/day)}}{\text{Flow (gallons/day)}}$$

Monitoring Anaerobic Zone SRT

In a basin where there is thorough and adequate mixing, the HRT and SRT should be similar for a given tank. However, as the solids concentration increases, and when mixing is minimal, a significant portion of the solids may settle in the basin. This will not only increase the SRT in the basin but will also increase the mixed liquor suspended solids (MLSS) concentration. This can result in insufficient carbon, excess VFAs, and even excess phosphorus release. SRT for the anaerobic zone is calculated as follows:

$$\text{SRT (days)} = \frac{\text{Anaerobic Zone Volume (mil gal)} \times \text{MLSS (mg/L)}}{\text{Waste Activated Sludge (WAS) Flow (mgd)} \times \text{WAS Concentration (mg/L)}}$$

Monitoring DO/ORP

It is absolutely critical that PAOs are cycled between anaerobic and aerobic conditions. To ensure this, the various zones must be monitored. An aerobic environment is characterized by the presence of free oxygen (O_2), whereas an anaerobic environment lacks free oxygen and, mostly, nitrate, but may still contain sulfate. For fermentation to work, even bound oxygen cannot be present. DO probes may not detect sulfates and sometimes not even nitrates.

ORP can be a more reliable method for monitoring zones to make sure that fermentation and the production of VFAs can take place (Table 9-3).

Monitoring Return Sludge Concentration

It is helpful to know the concentration of solids in the anaerobic zone. Remember, RAS is returned and mixed with influent wastewater. Although the RAS rate is set as a percentage of the flow, it is good to keep a handle on the actual solids concentration of the mixed RAS and influent. This will help determine how many microorganisms are available compared with the amount of carbon available. In other words, this will help determine the F/M ratio for the anaerobic zone. Parameters and monitoring locations in the anaerobic zone are shown in Table 9-4.

Monitoring Specific to the Aerobic Zone

Monitoring in the aerobic zone should show the following:
- a decrease in *ortho*-P from influent to effluent
- an increase in ORP (or DO) values from influent to effluent
- polyphosphate granules observed using the Neisser stain
- phosphorus content in the sludge greater than 5% dry weight basis

Monitoring parameters for the aerobic zone are shown in Table 9-5.

Table 9-3   Monitoring with ORP

ORP Range *mV*	Electron Acceptors	Conditions
−200 to −400	cBOD	Fermentation, anaerobic
−50 to −200	Sulfates	Fermentation, anaerobic
+50 to −50	Nitrates	Anoxic, anaerobic
+50 to +300	O2	Oxic, aerobic

cBOD—carbonaceous biological oxygen demand, ORP—oxidation–reduction potential

Table 9-4   Monitoring locations in the anaerobic zone

Parameter	Location
Carbon (COD or BOD)	Anaerobic zone influent
VFAs	Anaerobic zone effluent
TP	Anaerobic zone influent
Ortho-P	Anaerobic zone influent and effluent
pH	Anaerobic zone
ORP/DO	Anaerobic zone
HRT/SRT	Anaerobic zone
Solids concentration or F/M ratio	Anaerobic zone

BOD—biological oxygen demand, COD—chemical oxygen demand, DO—dissolved oxygen, F/M ratio—food-to-microorganism ratio, HRT—hydraulic retention time, ORP—oxidation–reduction potential, SRT—solids retention time, VFA—volatile fatty acid

Table 9-5   Parameters to monitor in the aerobic zone

Chemical/Biological Parameters	Process Control Parameters
• Carbon (BOD)	• HRT
• Carbon (COD)	• SRT
• Mixed liquor (MLSS)	• F/M ratio
• pH	• DO/ORP
• Phosphorus	• Nutrients

BOD—biological oxygen demand, COD—chemical oxygen demand, DO—dissolved oxygen, F/M ratio—food-to-microorganism ratio, HRT—hydraulic retention time, MLSS—mixed liquor suspended solids, ORP—oxidation–reduction potential, SRT—solids retention time, VFA—volatile fatty acid

Monitoring Aerobic Zone COD/BOD

One of the most valuable monitoring tools in activated sludge is the F/M ratio. Maintaining the proper F/M ratio in the aeration zone is key to achieving good performance in the treatment system. The F/M ratio is a measure of the load of "food" entering the zone compared the load of microorganisms in the zone. The loading to the *zone* is calculated using BOD or COD directly entering the zone.

Where to monitor BOD/COD.
• Direct influent to the aerobic zone

Monitoring Aerobic Zone pH

Floc-forming bacteria, along with other heterotrophic bacteria responsible for removing BOD in the aerobic zone, are also affected by pH, particularly unstable pH swings. It is important to monitor the pH of the influent into the aeration zone.

Where to monitor pH.
• Direct influent to the aerobic zone
• End of the aeration zone

Monitoring Aerobic Zone Phosphorus

Phosphorus entering the aerobic zone comes from two sources: influent to the treatment plant and phosphorus released in the anaerobic zone. The concentration of PAOs in the aeration basin will determine how much phosphorus uptake will occur. It will also alert the operator if too much phosphorus is being released in the anaerobic zone. Monitoring phosphorus entering the aerobic zone is important in ensuring that the MLSS concentration is sufficient.

It is well known the monitoring the F/M ratio (food entering the aerobic zone/microbes in the zone) is an important process control measure. However, monitoring just P (as food) compared with the microbes in the zone is equally important. Monitoring to establish the best P-to-MLSS or P-to-MLVSS ratio will help determine whether enough PAOs are being produced to meet the phosphorus load into the zone.

It is also helpful to monitor the phosphorus content of the settled activated sludge leaving the aerobic zone. Typically, the phosphorus content in activated sludge is approximately 1–3% but can increase to approximately 6–7% when EBPR is used. Generally, the EBPR system operates well if the phosphorus content in the activated sludge is greater than 5%.

Where to monitor phosphorus.
- Influent to the aerobic zone
- Settled activated sludge leaving the aerobic zone

Monitoring Aerobic Zone HRT and SRT

Generally, the HRT and the SRT can differ when measured in a secondary clarifier, particularly because the weir overflow rate for the liquid may be different from the rate at which the solids are being removed. In the aeration zone, however, the HRT and SRT are generally the same if the basin is well mixed with no short-circuiting.

Sufficient time must be allowed in the aeration zone to allow microorganisms to come in contact with and degrade the organic compounds and for PAOs to uptake and store phosphorus.

Monitoring Aerobic Zone F/M Ratio

The ratio of how much food (BOD) in pounds is available to the population of microorganisms (MLSS or MLVSS) in pounds is critical to proper operations. The goal is to match the amount of food with the number of microorganisms. Treatment system upsets can result when there is too much food and not enough microorganisms or when there are too many microorganisms and not enough food. The F/M ratio is calculated as follows:

$$\text{F/M Ratio} = \frac{\text{Daily Mass of Food (lb)}}{\text{Daily Mass of Microorganisms (lb)}}$$

Pounds of Food = Influent BOD (mg/L) × Flow (mil gal) × 8.34
Pounds of Microorganisms = MLVSS × Tank Volume × 8.34

Table 9-6   Monitoring locations in the aerobic zone

Parameter	Location
Carbon (COD or BOD)	Direct influent to aerobic zone
pH	Direct influent to aerobic zone End of aerobic tank
TP	Direct influent to aerobic zone Settled mixed liquor leaving aerobic tank
ORP/DO	Beginning and end of aerobic zone
HRT/SRT	Aerobic zone
Solids concentration or F/M ratio	Aerobic zone

BOD—biological oxygen demand, COD—chemical oxygen demand, DO—dissolved oxygen, F/M ratio—food-to-microorganism ratio, HRT—hydraulic retention time, ORP—oxidation-reduction potential, SRT—solids retention time, TP—total P

The F/M ratio should be monitored for each aeration basin in service. The pounds of food should be calculated based on the direct influent to the basin, as well as the actual concentration of microorganisms in that basin.

Monitoring Aerobic Zone ORP or DO

Aerobic bacteria each require 0.1–0.3 mg/L of oxygen to function properly. However, a residual of 0.8–2.0 mg/L of DO should be maintained for effective treatment. Floc particles are made up of mostly bacteria. So, enough DO must be available to penetrate the floc. Otherwise, the bacteria in the center of the floc will die, floc solids will begin to break apart, and unfavorable filamentous bacteria that thrive in low-DO conditions can begin to develop in the floc. Additionally, having residual DO assures reduction of BOD, nitrogen compounds, and phosphorus. The ORP measurement in the aerobic zone should range from +50 to +300. Monitoring parameters and locations are shown in Table 9-6.

Microbiological Monitoring in the Anaerobic Zone

PAOs in the Anaerobic Zone

In the anaerobic zone, PAOs absorb VFAs and store them in the cell as an insoluble starch, i.e., PHB. PHB can be stored inside the cell and can even be attached to the outside of the cell body as discrete granules. The presence of PHB granules in the PAO cell is an indication that sufficient VFAs were present in the anaerobic zone and that phosphorus release has occurred. The abundance of PHB measured in the anaerobic zone can be used as an indicator of the conditions in the anaerobic zone and as an indicator of the subsequent uptake of phosphorus in the aerobic zone.

PAOs are morphologically diverse. Species appear as round, oval, or rod-shaped cells. PAOs appear as individual cells early in the process but will grow in clusters as the PAO population increases. Stored PHB granuals cannot be visualized with conventional microscopy. PHB is a lipid-like polymeric ester

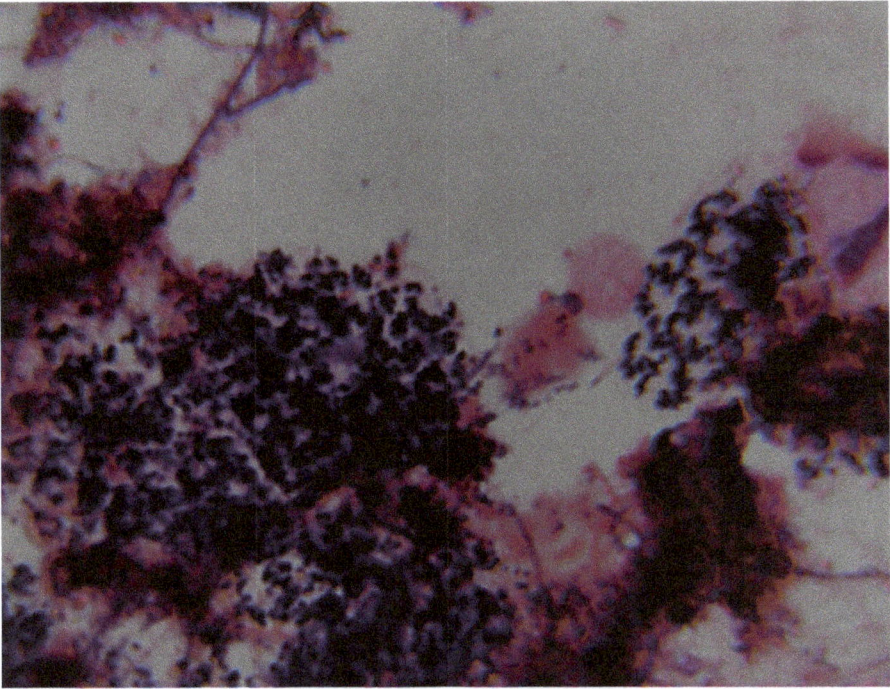

Figure 9-8 Sudan Black–stained PHB inclusions in PAO cells collected in the anaerobic zone (Sudan Black stained and observed at 100× magnification with oil immersion, bright-field)

stored in and on the cell as descrete inclusions or granules. These granules are not visible without staining. The Sudan Black (PHB) staining method can be used to evaluate the abundance of PAOs in a sample. Sudan Black is a lipophilic stain with the ability to dissolve in fatty lipid material. This allows for the detection of PHB granules, which would otherwise not be visible under light microscopy. Because this stain is specific to PHB, it should only be used on samples collected from the anaerobic zone. Stored PHB in the PAO cell appears as intracellular blue-black granules in and around the cell (Figure 9-8). PAOs observed in the anaerobic zone are designated as PAO–PHB.

GAOs in the Anaerobic Zone

GAOs are relatively large, hollow-looking cells that generally grow in clusters and can also appear as large groups of tetrads.

In the anaerobic zone, GAOs compete with PAOs for the VFAs. GAOs use their stored intracellular glycogen as an energy source for storing PHB. GAOs are large cells with the ability to take up VFAs under anaerobic conditions but are unable to take up phosphorus under the subsequent aerobic conditions. In other words, they take up VFAs but do not contribute to phosphorus removal. Increasing numbers of GAOs are an indication of performance problems in the EBPR system.

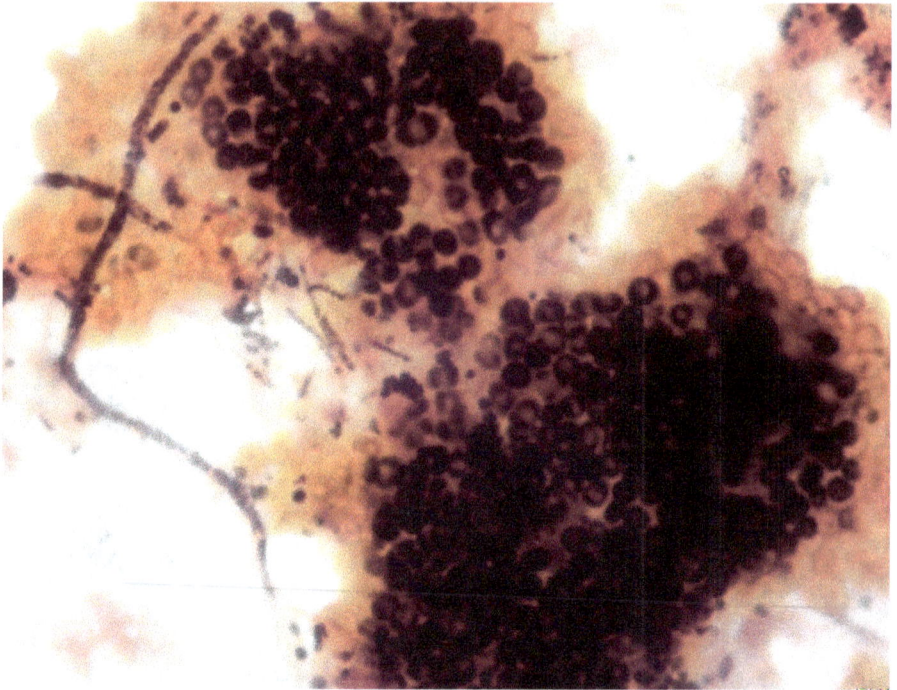

Figure 9-9 Sudan Black–stained PHB inclusions (100× magnification with oil immersion, bright-field)

GAOs in the anaerobic zone can also be observed using the Sudan Black stain. Both PAOs and GAOs store PHB and can stain PHB-positive with the PHB staining procedure (Figure 9-9). There are morphological differences between PAOs and GAOs. GAOs can appear as large, hollow-looking cells while PAOs and smaller and more dense. But, because both stain positive, this can make it difficult to distinguish between the tow in the anaerobic zone. Stored PHB in the GAO cell appears as intracellular blue-black granules in and around the cell.

Microbiological Monitoring in the Aerobic Zone

PAOs in the Aerobic Zone

Under aerobic conditions, PAOs use the energy gained from the breakdown of stored PHB for the uptake of phosphorus. Phosphorus is stored in the cell in the form of polyphosphate granules (poly-P). Stored poly-P granules can be visualized using the Neisser stain procedure. Like PHB, poly-P granules are not visible with staining. Methylene blue in the Neisser stain gives polymeric poly-P chains a purple-black color. The characteristic positive reaction of the Neisser staining method is a purple-black granule in a yellowish-brown background of the counterstained cells (Figure 9-10).

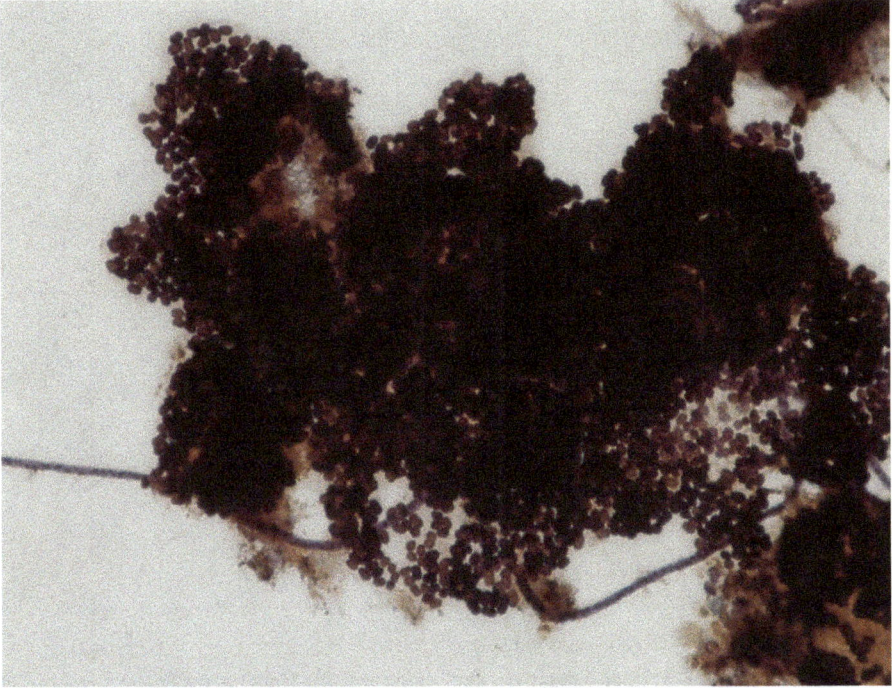

Figure 9-10   Neisser-stained polyphosphate granules (Neisser stained and observed at 100× magnification with oil immersion, bright-field)

The Neisser stain is specific to stored poly-P granules, so it should only be used with samples collected from the aerobic zone of the EBPR process. PAOs observed in the aerobic zone are designated as PAO–poly-P.

GAOs in the Aerobic Zone

In the aerobic zone, GAOs metabolize PHB to gain energy for glycogen production and do not uptake phosphorus. GAOs in the aerobic zone do not stain positive for the Neisser stain. They appear as hollow cells that are brown to orange and can be observed with either the Neisser or Gram stains (Figure 9-11). The complete Sudan Black (PHB) and Neisser staining procedures are described in Chapter 3.

Because the EBPR process depends on the metabolism of PAO communities, monitoring both the PAO–PHB (anaerobic zone) and the PAO–poly-P (aerobic zone) communities, as well as the GAO communities in the EBPR system, can provide important guidance for stable operations. One way to do this is to compare the abundance of PAO–PHB with the abundance of PAO–poly-P.

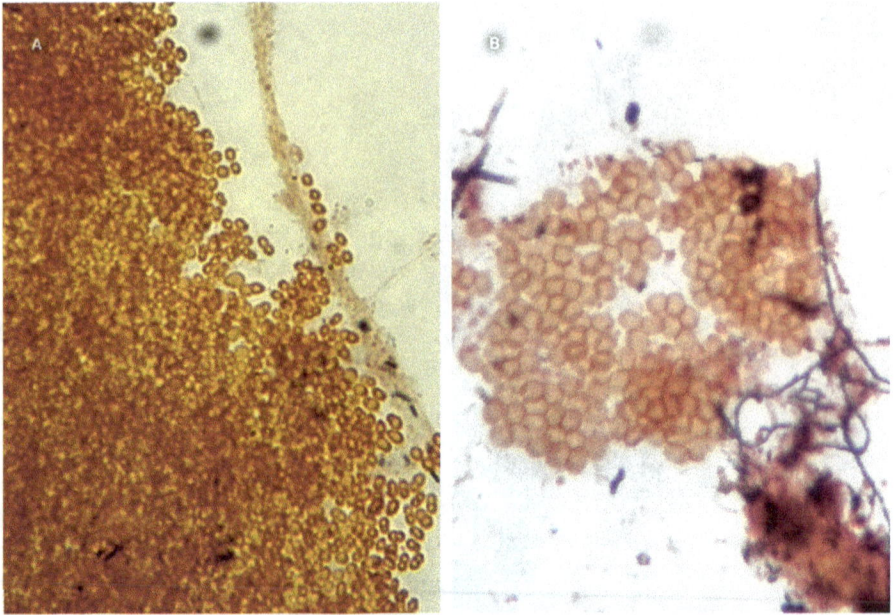

Source: Image B printed with permission from Glymph-Martin 2024

Figure 9-11 Neisser stained negative and observed at 100× magnification using oil immersion, bright-field microscope. (A) and (B) Neisser-stained GAOs in the aerobic zone

The abundance of PHB granules observed in the anaerobic zone is an indicator of VFA uptake and the subsequent storage of PHB in the cell. PAO-PHB abundance in the anaerobic zone is directly related to PAO–poly-P abundance and phosphorus uptake in the aerobic zone. However, studies have shown that PHB can be produced by more than 90 genera of gram-positive and gram-negative bacteria under both aerobic and anaerobic conditions using several carbon sources. This includes GAOs. Consequently, a high PAO–PHB abundance in the anaerobic zone may not be predictive of a high corresponding PAO–poly-P abundance (phosphorus uptake) in the aerobic zone. A high PAO–PHB abundance in the anaerobic zone with a corresponding low PAO–poly-P abundance in the aerobic zone suggests that a portion of the PHB uptake in the anaerobic zone is accomplished by microorganisms other than PAOs. Therefore, if you see a lot of PHB and only a little poly-P, this is an indication that there may be issues in the anaerobic zone and that the environment is not favorable to only PAOs.

Most of all, it is helpful to regularly observe samples from the anaerobic zone and the aerobic zone, particularly just to make sure that there is a healthy population of PAOs and to look for an abundance of GAOs so that the proper adjustments can be made.

REFERENCES

Gerardi, M.H. 2006. *Wastewater Bacteria*. Hoboken, N.J.: Wiley-Interscience.

Glymph-Martin, T. 2024. *Activated Sludge Microbes Poster 2*. Wastewater Microbiology Solutions.

Pasztor, I., P. Thury, and J. Pulai. 2009. "Chemical Oxygen Demand Fractions of Municipal Wastewater for Modeling of Wastewater Treatment." *International Journal of Environmental Science and Technology*. 6:51-56.

Pattarkine, V.M. and C.W. Randall. 1999. "The Requirement of Metal Cations for Enhanced Biological Phosphorus Removal by Activated Sludge." *Water Science and Technology*. 40(2):159-165.

Chapter 10

Biological Nutrient Removal: Ammonia

Nitrogen is an essential nutrient for all living organisms because it is a fundamental component of vital molecules in the body, along with amino acids, proteins, DNA, and RNA. Nitrogen compounds are found in fecal and food waste that are discharged to the wastewater treatment plant. Nitrogen enters the treatment plant in both organic and inorganic forms. Organic nitrogen can be found bound to organic molecules such as amino acids, proteins, and building blocks of DNA and RNA. A significant source of organic nitrogen is urea, which is a major component of urine. Approximately 60% of nitrogenous waste entering the treatment plants is organic, and 40% is inorganic.

When organic nitrogen compounds such as amino acids and urea enter the sewer system, inorganic nitrogen compounds are formed. Ammonia and ammonium (i.e., ionized ammonia) are two forms of inorganic nitrogen that are formed. Ammonia is toxic and is mostly released to the atmosphere from the sewer system or biological treatment through aeration or mixing action. Ionized ammonia is nontoxic and is used by bacteria as their preferred source of nutrient nitrogen. The quantity of each of the two forms present in the wastewater will depend on the pH. At pH values less than 9.0, most of the inorganic nitrogen present is ionized ammonia. The ionized ammonia concentration in typical municipal wastewater is 25–30 mg/L.

When a high concentration of ammonia is discharged into the waterways, it can be toxic to fish and other aquatic organisms. Ammonia can also exert oxygen (O_2) demand on the receiving water. When present, bacteria and other microorganisms in the water oxidize ammonia into nitrite and nitrate. This process consumes oxygen, diminishing oxygen levels in the receiving stream and causing fish kill. Excess ammonia, as well as phosphorus (P), can also contribute to nutrient loading in waterbodies. This can cause plant overgrowth and lead to harmful algal blooms. Therefore, managing ammonia is crucial for maintaining healthy aquatic ecosystems.

NITRIFICATION

Nitrogen is removed or reduced in wastewater through the process of *nitrification*. Nitrification in activated sludge treatment systems can be achieved in one or two stages. In a one-stage process, aeration basins are designed to remove both carbonaceous biochemical oxygen demand (cBOD) and nitrogenous BOD (nBOD). In a two-stage process, stage 1 aeration basins are designed to remove cBOD, and stage 2 aeration basins are designed to nitrify or remove nBOD. cBOD refers to the oxygen demand created when bacteria break down organic compounds in the wastewater, whereas nBOD refers to the oxygen demand created by the oxidation of nitrogen-containing compounds such as ammonia.

Nitrification is a two-step process carried out by different types of bacteria. It is the biological oxidation of ionized ammonia to nitrite and/or nitrite to nitrate. For the remainder of this chapter, when I speak of ammonia, I will be in reference to ionized ammonia (i.e., ammonium).

There are six major groups of nitrifying bacteria involved in the nitrification process. Ammonia-oxidizing bacteria (AOB) oxidize ammonia to nitrite. There are three major groups of AOB:

- *Nitrosomonas*
- *Nitrosococcus*
- *Nitrosospira*

Nitrite-oxidizing bacteria (NOB) oxidize nitrite to nitrate. There are three major groups of NOB:

- *Nitrobacter*
- *Nitrospira*
- *Nitrococcus*

Nitrifying bacteria (nitrifiers) are free-living microorganisms that live in soil and water (Figure 10-1). They enter the wastewater treatment system naturally in the wastewater and through inflow and infiltration. Nitrifiers are *chemolithoautotrophs*, meaning that they get their energy from inorganic compounds—in this case, ammonia.

Nitrifiers have long, thin structures inside the cell called *cytomembranes* that use electrons from the nitrogen atom in ammonia to produce energy. Because only a small amount of energy is gained when nitrifiers convert ammonia to nitrite and nitrite to nitrate, nitrifiers produce very slowly, and sludge production is relatively small. Generation times can range from 8 to 10 hours compared with 20 min for the floc-forming bacteria, and that's under optimal conditions. However, the activated sludge environment is not the most ideal for nitrifiers, so the generation time can be extended to as long as two to three days. Therefore, a long mean cell residence time is required to establish a sufficient population of nitrifiers. Nitrifiers are not floc formers. However, they do easily adsorb to floc particles. Because they require significant amounts of oxygen, they can be found attached along the outer edges of the floc.

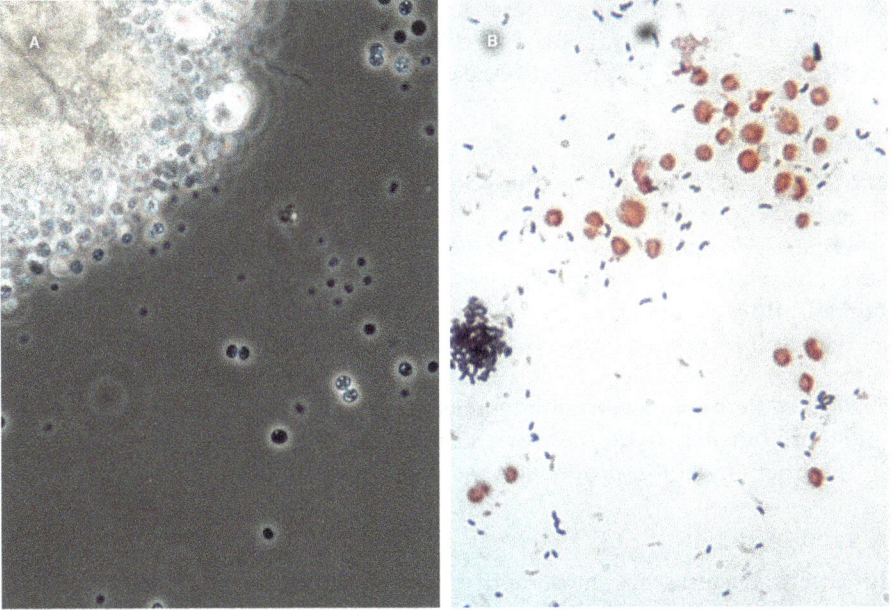

Figure 10-1 *Nitrosococcus*: 100× magnification, oil immersion, phase contrast (A); and 100× magnification, oil immersion, Gram stain, bright-field (B)

FACTORS INFLUENCING THE GROWTH AND ACTIVITY OF NITRIFYING BACTERIA

There are several factors that can influence the growth and the activity of nitrifiers in the treatment system. These include dissolved oxygen (DO), temperature, aeration basin solids retention time (SRT), alkalinity, pH, and toxicity.

Dissolved Oxygen

Nitrifying bacteria are strict aerobes, meaning that nitrification can only occur in the presence of free molecular oxygen (O_2). For every milligram of ammonia oxidized to nitrite, approximately 4.6 mg of oxygen is consumed. Nitrification generally improves as DO concentration increases up to a maximum nitrification rate at approximately 3.0 mg/L of DO. However, nitrification will diminish significantly at DO concentrations less than 0.5 mg/L.

Temperature

The rate of nitrification is temperature dependent. Like most bacteria, nitrifier activity and reproduction increase with rising temperatures. Although nitrifiers can reproduce in temperatures ranging from 5 to 40°C, the maximum temperature for nitrification is approximately 30°C. Nitrification activity will decline by 50% at 15°C and will cease at temperatures less than 5°C.

Aeration Basin SRT

Nitrifiers are strict aerobes; therefore, the nitrification process takes place only in the presence of oxygen. Even under ideal conditions in the activated sludge treatment system, the generation time for nitrifiers can take several days. So, naturally, a longer aeration basin SRT is required to maintain a healthy population of nitrifiers. As temperature decreases and nitrifier activity and reproduction decrease, a longer SRT will be required. Not only does a longer SRT allow more time for growth and reproduction, but also the mixed liquor concentration increases, providing more time in the system and more nitrifiers to do the work. Conversely, as temperatures increase and nitrifier activity and reproduction increase, SRT can be reduced.

Alkalinity and pH

Nitrifying bacteria are most active at pH values between 6.8 and 7.2. Nitrification decreases significantly at pH values less than 6.5 and greater than 8.0. When ammonia is oxidized to nitrite, hydrogen ions are released. When hydrogen ions are released in solution, the pH will drop. If alkalinity is not present, the pH will continue to decline. Once the pH reaches 6.5, nitrification will decrease significantly. For every one part of ammonia oxidized, 7.14 parts of alkalinity are consumed. Therefore, adequate alkalinity is essential for successful nitrification.

Toxicity

Nitrifiers are dependent on inorganic carbon for their carbon source. Their ability to oxidize (break down) ammonia and nitrite can be inhibited in the presence of short-chain alcohols (i.e., methanol and ethanol) or amines (i.e., methylamine and ethylamine). These short-chain compounds hinder the nitrifier's enzymes and inhibit their ability to oxidize ammonia and nitrite. Nitrifiers are also affected by other toxic chemicals entering the treatment system.

Although ammonia and nitrite are energy substrates for nitrifiers, ammonia can still be toxic at concentrations greater than 480 mg/L (Gerardi 2006). It seems strange that the very thing nitrifiers need for energy could end up being toxic for them. However, it is not the ammonia (ionized) itself that is toxic; rather, the process of converting ammonia to nitrite is much faster than the process of converting nitrite to nitrate. Therefore, excess ammonia will create excess nitrite. The oxidation of nitrite to nitrate requires oxygen, so excess nitrite will also exert an oxygen demand. In addition, excess nitrite can damage nitrifiers, leading to stress or cell death. Even worse, at decreasing pH levels, excess nitrite can be converted to nitrous acid in water, and nitrous acid is more toxic than nitrite itself and will inhibit microbial activity.

DENITRIFICATION

Denitrification is a critical process for removing nitrogen from wastewater by converting nitrate into nitrogen gas. The nitrogen gas produced is inert and escapes harmlessly into the atmosphere. Denitrification is performed by facultative anaerobic bacteria under anoxic (low-oxygen) conditions. These bacteria use oxygen bound to nitrate as an alternative to free molecular (unbound) oxygen for respiration. However, cBOD must also be present. In other words, they only need oxygen if food is present.

Generally, in activated sludge systems that are required to remove nitrogen, an anoxic zone is present where denitrification can occur. However, in systems designed to reduce ammonia, sometimes denitrification can occur in secondary clarifiers. In secondary clarifiers, solids settle to the bottom and form a sludge blanket. If sludge is retained in the clarifier for too long, microorganisms consume available oxygen in the sludge layer. This will promote the development of anoxic conditions within this sludge. This low-oxygen condition is ideal for denitrifying bacteria to use nitrate for respiration. Remember, nitrate is produced in the aeration tanks during nitrification (conversion of ammonia to nitrate). If wastewater entering the clarifier still contains significant levels of nitrate, denitrification can occur in the anoxic areas in the sludge. In addition, if sludge is held too long in the clarifiers, this increases the likelihood of anoxic conditions and denitrification.

Denitrification produces nitrogen gas bubbles. These bubbles can attach to sludge particles and cause the sludge to rise to the surface. This can disrupt settling, reduce clarifier efficiency, and lead to solids washout and increased total suspended solids.

How do you prevent denitrification in the clarifiers?
- Regular and efficient sludge removal: To prevent denitrification in the clarifiers, it is important to ensure regular and efficient sludge removal to minimize sludge blank depth and retention time.
- Control nitrate levels: Reduce nitrate carryover from the aeration tank by optimizing aeration and ensuring appropriate anoxic zones for denitrification upstream.
- Monitor sludge blanket levels: Regularly measure sludge blanket depth to prevent excessive accumulation.
- Adjust return-activated sludge (RAS) rates: Maintain optimal RAS flow rates to prevent sludge buildup and maintain a proper balance in the system.
- Proper clarifier design: Use mechanical mixers or other techniques to minimize sludge accumulation in the clarifier and to avoid dead zones where sludge may stagnate.

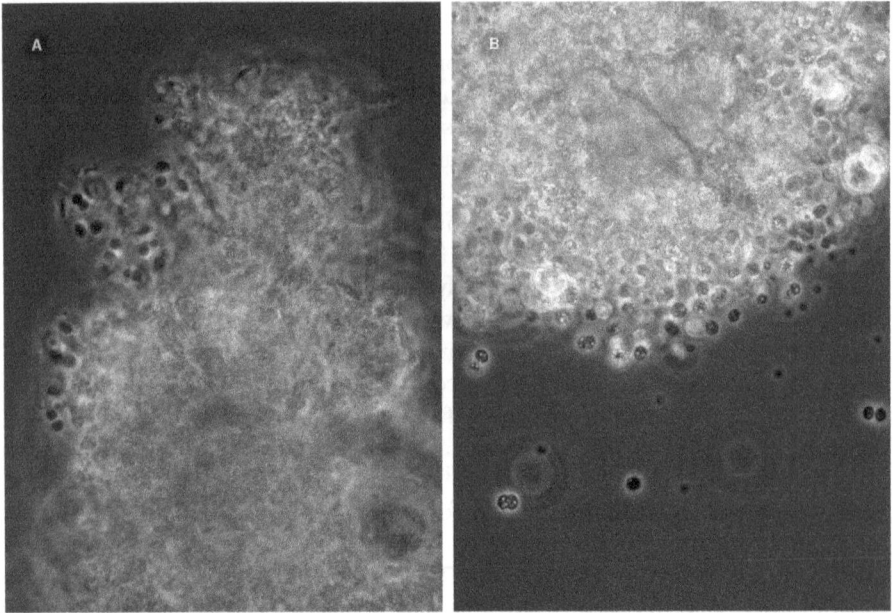

Figure 10-2 Nitrifying bacteria attached to floc particles (100× magnification, oil immersion, phase contrast) (A and B)

NITRIFIERS UNDER THE MICROSCOPE

Nitrifying bacteria are difficult to see using traditional microscopy. It is even more difficult to distinguish the different species. There are more advanced fluorescence probe methods, but that would defeat the purpose of keeping it simple. However, they can be observed using the 100× oil-immersion objective with phase contrast. Nitrifiers are naturally attracted to floc particles, so you can find them along the edges of the floc if you look very closely (Figure 10-2).

Microbiological monitoring of nitrification systems is not simple. The best strategy for these systems is to monitor the conditions that affect nitrifiers, such as temperature, alkalinity, pH, aeration basin detention time, and toxicity.

REFERENCES

Gerardi, M.H. 2006. *Wastewater Bacteria*. Hoboken, N.J.: Wiley.

Glymph-Martin, T. 2024. *Activated Sludge Microbes Poster 2*. Wastewater Microbiology Solutions.

Chapter 11

Microbiology and Process Control

It is beneficial to develop and maintain a process control program that includes microbiology. Conducting simple analyses routinely and consistently is key to understanding the microbial population of your wastewater treatment plant. Becoming familiar with what is normal for your plant and monitoring trends can be an addition to your process control program. It is important, however, to try to be as consistent as possible. In other words, always collect the sample from the same location, and always use the same number of drops on the slide.

Considering the time restraints of most operators, sometimes counting protozoa and metazoan species or identifying filamentous bacteria is not feasible. Therefore, there are other microbiological indicators that don't require a lot of time that can help the operator evaluate treatment system conditions. These include looking at the following:

- Floc characteristics
- Surrounding fluid
- Health and activity of protozoa/metazoa
- Nutrient deficiency

WHAT CAN THE FLOC TELL US?

The development of healthy floc is critical for good separation of the solids from the liquid treated water. There are several ways to tell whether the floc is healthy.

Floc Color

Under normal conditions, floc should appear brown. This indicates that sufficient food/nutrients are present or that the system is operating at a sufficient food-to-microorganism (F/M) ratio. It also means that there is no stress to the bacteria. When bacteria are starved of food or when nutrients are deficient, bacteria produce excess amounts of exocellular lipopolysaccharide. Bacteria will also produce excess amounts of lipopolysaccharide in the presence of excessive amounts of organics acids or adverse conditions such as toxicity or pH issues. Organic acids are introduced in the system through anaerobic recycle streams or extremely low–dissolved oxygen (DO) conditions. The presence of excess amounts of exocellular lipopolysaccharide will give the floc a milky white color and slimy consistency. Exocellular lipopolysaccharide is viscous and slimy and can cause bulking and foaming in the treatment system.

A simple sludge color index (SCI) can be used to rank the color of the sludge floc. The index ranges from –5 (white-color floc) to +5 (black-color floc), with the target rank of "0" used to represent a healthy brown color. The more negative the number, the more lipopolysaccharide (the lighter the floc color). A more positive number represents older or more septic sludge (darker in color) (Figure 11-1).

Floc Size and Density

Big is not necessarily better. Floc particles are made up of bacteria that have amassed together to form floc. If floc particles are too large (>500 μm) (Glymph 2011), it will be difficult for DO and organic substrate to penetrate to the center of the floc. On the other hand, floc particles that are too small may not settle well. Ideally, floc should measure 200–300 μm. In addition, if floc is too dense, it will settle rapidly, leaving behind a turbid effluent.

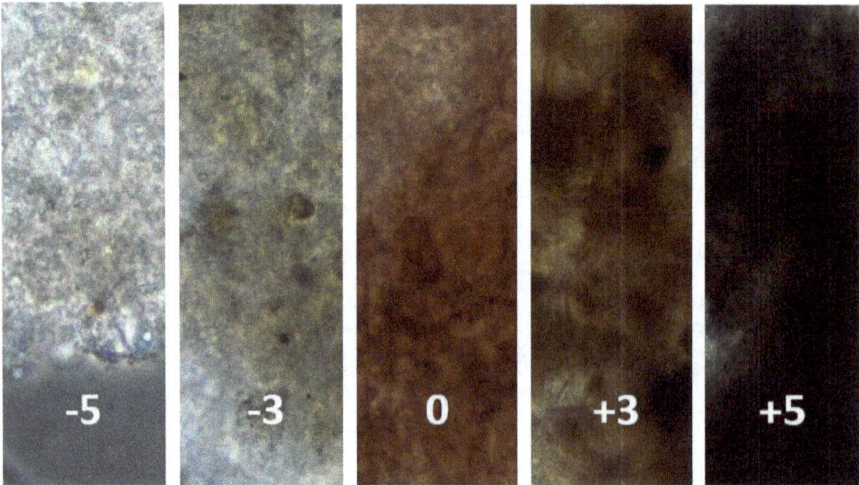

Source: Printed with permission from Glymph 2011

Figure 11-1 Sludge color chart

WHAT CAN THE SURROUNDING FLUID TELL US?

Dispersed Bacteria

Under normal conditions, when samples are collected from the discharge end of the aeration basin, the fluid surrounding floc particles should be relatively free from dispersed bacteria. This is because healthy, active protozoa and metazoans do a good job of removing excess and nonflocculating bacteria from the effluent. It also means that floc-forming bacteria are healthy and forming floc. When significant amounts of dispersed bacteria are present in the fluid surrounding the floc, this is an indication that very few ciliates are present or that something is interfering with floc formation. Excessive amounts of dispersed bacteria can also cause an increase in effluent turbidity (Figure 11-2).

Zooglea

There are basically two types of bacteria in activated sludge: those that form floc and those that do not. Under adverse conditions, floc formers produce excess amounts of exocellular polymers, which is evident by the milky white color of the floc. However, non–floc formers surround themselves in exocellular polymers to form zoogleal communities that serve as protection

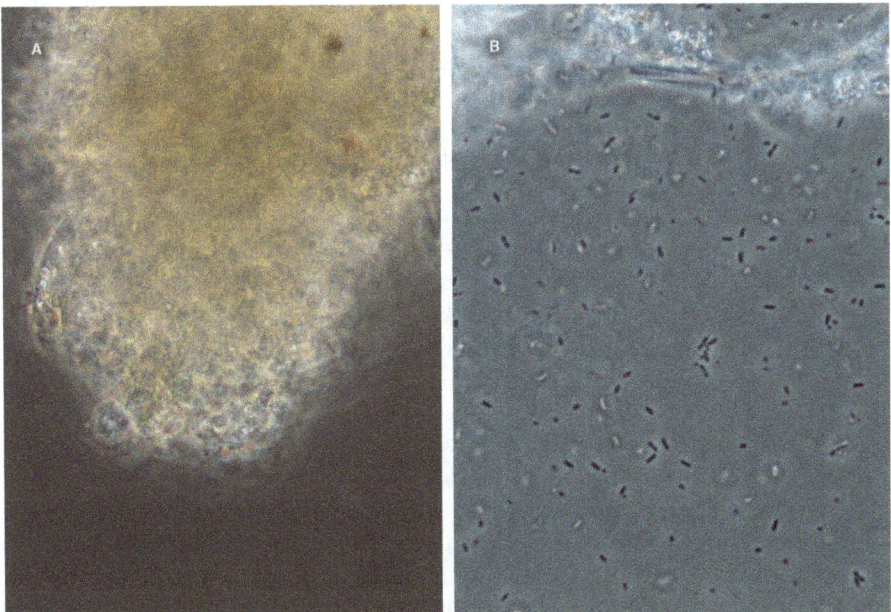

Source: Image B printed with permission from Glymph-Martin 2024

Figure 11-2   Clear fluid surrounding the floc (A) and excess dispersed bacteria in the fluid (B)

against adverse conditions. These masses can be seen floating freely in the bulk fluid or attached to the edges of floc particles. They can appear as a fingerlike projection and irregular-shaped masses. Sometimes, these zoogleal masses will form within the existing floc particle, forming very dense zoogleal mass floc particles (Figure 11-3).

These zoogleal masses are viscous and can cause increases in effluent turbidity and can interfere with sludge settling. Adverse conditions can range from starvation (extremely low F/M), nutrient deficiency, changes in pH, or toxicity. Some zoogleal masses can be normal. This simply means that something in the environment has caused this protective response from the bacteria. This can be low levels of toxicity from industry and wash from stormwater, among other things. Their presence can also mean that there is not enough food or nutrients available. It simply means the bacteria are stressed. However, increasing numbers of these masses can be an indication that something toxic has entered the treatment system.

HEALTH AND ACTIVITY OF PROTOZOA/METAZOA

In addition to looking at the floc or the surrounding fluid, monitoring the dominance of the different protozoa and metazoan species can also give us an indication of treatment system conditions. Under normal conditions, protozoa and metazoans should be prominent in the activated sludge system. A healthy population of ciliates (stalked, crawling, free-swimming), as well as some amoebae, flagellates, and metazoans such as rotifers and nematodes, may also be observed. I like to put the protozoa and metazoa in four groups:

- Group 1: amoebae and flagellates
- Group 2: all ciliates (free-swimming, crawling, and sessile or stalked)
- Group 3: all metazoa
- Group 4: all shelled species (shelled amoebae, shelled rotifers, and tube-dwellers)

Generally, if the system is operating properly, group 2 should dominate. In other words, there should be more ciliates than any of the other groups. If group 1 is dominant, the system is "young," and more detention time is required in the aeration basin. If group 3 is dominant, the system is "old," and less detention time is required in the aeration basin. However, under adverse conditions, protozoa and metazoans that have the ability to develop protective mechanisms will begin to dominate in the system. Shelled amoebae, shelled rotifers, and tube-dwelling ciliates such as *Vaginicola* have the ability to form protective shells or tubes that enable them to survive in adverse conditions. In municipal activated sludge systems, generally, shelled species account for about 25% of the total protozoa and metazoans present. However, in industrial systems, or municipal systems with significant industrial input, shelled species can account for 40–50% of the total. If you don't have time to do a full protozoa/metazoa count, it will be much easier just to track shelled species. If the number increases significantly and they

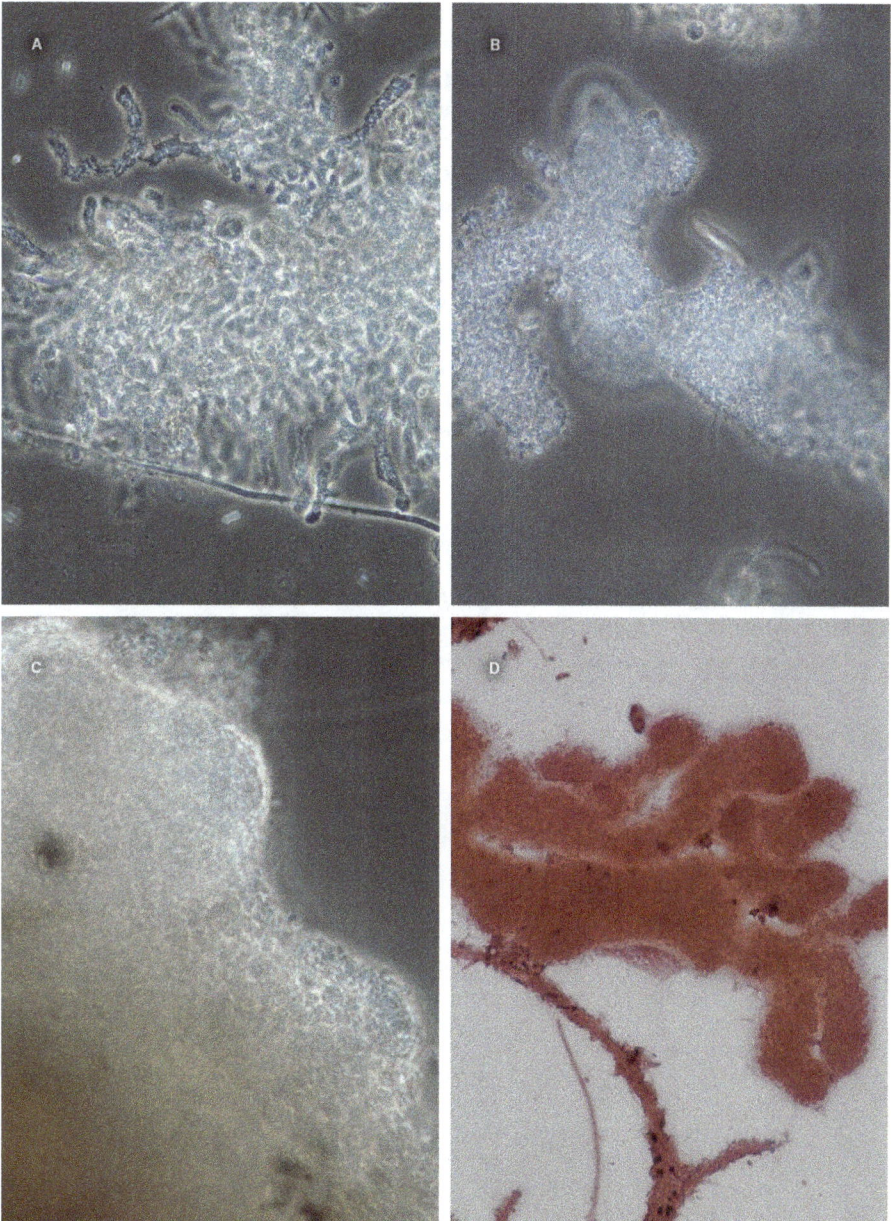

Figure 11-3   Zoogleal mass with fingerlike projections (40× magnification, phase contrast) (A); zooglea mass (40× magnification, phase contrast) (B); dense zoogleal mass floc particle (100× magnification, oil immersion, phase contrast) (C); and zoogleal mass with fingerlike projections (40× magnification, Gram stained, bright-field) (D)

begin to dominate, this is an indication that treatment system conditions are negatively affecting the microorganisms. If both zooglea masses and shelled species increase significantly, this can be an indication that substances toxic to the microoganisms have entered the treatment system.

Monitoring your treatment system microbiology doesn't have to be complicated. If all you have time to do is look at the floc color, then look at the floc color on a regular basis. If all you have time to do is look at the fluid surrounding the floc, then just do it on a regular basis. Most importantly, keep a process chart of treatment process parameters so you can compare what the treatment system microbiology looks like when the system is working well. The more you do it, the better you will be at it, and the more equipped you will be to monitor treatment system performance, predict upsets, and make the needed adjustments to keep the microbes happy. If you treat the microbes right, they will treat the water right.

REFERENCES

Glymph, T. 2011. *A Wastewater Microbiology Laboratory Manual for Operators.* Operator Training Committee of Ohio (OTCO): Columbus.

Glymph-Martin, T. 2024. *Activated Sludge Microbes Poster 2.* Wastewater Microbiology Solutions.

Glossary

aerobic   Refers to organisms or processes that require oxygen to survive or function.

anaerobic   Refers to organisms or processes that do not require oxygen for survival or function.

brightfield illumination   The simplest of all the optical microscopy illumination techniques. Sample illumination is transmitted by white light.

cryptobiosis   A biological state in which an organism enters an extreme form of dormancy to survive unfavorable environmental conditions, such as desiccation, freezing, oxygen deprivation, or high salinity.

darkfield illumination   A microscopy technique in which the specimen is illuminated with light that does not enter the objective lens directly. Instead, only light scattered by the specimen is captured, making the background appear dark and the specimen bright.

denitrification   A biological process in which nitrates are reduced to nitrogen gas or nitrous oxide by bacteria, releasing it into the atmosphere. This process occurs in oxygen-depleted environments.

differential interference contrast (DIC)   A high-resolution microscopy technique that produces a three-dimensional, shadowed effect.

differential staining   A staining technique that uses multiple dyes or steps to distinguish between different types of cells, structures, or cellular components.

facultative   Facultative organisms can grow in both the presence and absence of oxygen.

fermentation   A metabolic process that converts sugars into simpler compounds, such as alcohol or lactic acid, in the absence of oxygen.

glycogen-accumulating microorganisms (GAOs)   Microorganisms that hydrolyze glycogen to gain energy and may compete with PAOs for volatile fatty acids in the biological phosphorus removal process.

hydrolysis   A chemical reaction in which complex molecules are broken down into smaller soluble units.

mixed liquor A mixture of return activated sludge primary effluent combined and aerated in the aeration basin.

mixed liquor suspended solids (MLSS) The concentration of suspended solids in an aeration tank during the activated sludge process. The unit MLSS is primarily measured in milligrams per liter (mg/L).

mixed liquor volatile suspended solids (MLVSS) The amount (mg/L) of organic or volatile suspended solids in the mixed liquor of an aeration tank. This volatile portion is used as a measure or indication of the microorganisms present.

nitrification A biological process in which ammonia or ammonium is oxidized to nitrites and then to nitrates by certain bacteria.

oil immersion A technique used to increase the resolving power of a microscope. This is achieved by immersing both the objective lens and the specimen in a transparent oil of high refractive index, thereby increasing the numerical aperture of the objective lens.

phase contrast A microscopy technique that uses the differences in the phase of light transmitted or reflected by a specimen to form distinct, contrasting images of different parts of the specimen.

phosphorus-accumulating microorganisms (PAO) Microorganisms involved in biological phosphorus removal and capable of the uptake of excess amounts of phosphorus.

polyhydroxybutyrate (PHB) A polyhydroxyalkanoate (PHA) stored as discreet granules in the cells of PAOs in the presence of volatile fatty acids during anaerobic conditions.

polyphosphate During enhanced biological phosphorus removal, phosphorus from the aerobic zone is stored in the PAO cell in the form of polyphosphate granules.

preliminary treatment The first step in wastewater treatment, focusing on removing large solids, grit, and debris to protect downstream equipment and improve overall process efficiency. It typically involves screening, grit removal, and flow equalization.

primary sedimentation Also called primary clarification, this process involves the removal of settleable organic and inorganic solids from wastewater by gravity. The water flows into large sedimentation tanks, where suspended particles settle at the bottom as sludge and lighter materials, such as grease, float to the surface for removal.

saprophytic Saprophytic organisms obtain nutrients by decomposing and absorbing organic matter from dead or decaying plants and animals.

secondary treatment A biological process aimed at removing dissolved and suspended organic matter from wastewater. It typically uses microorganisms to break down organic pollutants in aeration tanks or biofilters.

shelled (testate) Some protozoa and metazoans produce shells (tests) by either secreting them or accreting them from appropriately sized particles encountered in the surrounding environment.

simple staining A staining technique that uses a single dye to color cells or structures, allowing for basic visualization of their shape, size, and arrangement under a microscope.

zooglea A colony of bacteria embedded in the polymeric or gelatinous mass.

Index

Note: *f.* indicates figure; *t.* indicates table

A

about phosphorus, 106
activated sludge, 5, 30, 54
activated sludge microbes, 11*f*
activated sludge process, 6
activated sludge treatment, 29
adaptation period, 28
advanced microscopes, 8
aeration basin, 11, 30
amoebae, 38
aquatic worms, 101

B

bacillus-shaped bacteria, 22, 23*f*
bacteria, 21
bacteria of different shapes, 22*f*
bacterium species, 31
beggiatoa, 70*f*
Beggiatoa with sulfur granules, 91*f*
biochemical oxygen demand, 29
biochemical reactions, 26
biological activity, 6
biological nutrient removal, 133
biological phosphorus removal, 105
biosolids, 5
brightfield illumination, 8, 10*f*
bright-field microscope, 130*f*
bristle worm, 65*f*
bulk fluid, 5

www.ingramcontent.com/pod-product-compliance
Lightning Source LLC
Chambersburg PA
CBHW070725220326
41598CB00024BA/3309